M000250473

SRA
Connecting Math Concepts

Level F Textbook

COMPREHENSIVE EDITION

A DIRECT INSTRUCTION PROGRAM

McGraw Hill Education

Bothell, WA • Chicago, IL • Columbus, OH • New York, NY

MHEonline.com

 **Education**

Send all inquiries to:
McGraw-Hill Education
8787 Orion Place
Columbus, OH 43240

ISBN: 978-0-02-103645-5
MHID: 0-02-103645-4

Printed in the United States of America.

3 4 5 6 7 8 9 QVS 15 14

Lesson

Everything that is introduced in this program is something you will learn. It is something that you'll need in order to work difficult problems that will be introduced later. Listen very carefully to the directions and follow them. Sometimes your teacher will direct you to work **part** of a problem, sometimes a **whole** problem, and sometimes a **group** of problems. Work quickly and accurately. Most important, work hard.

Part 1 Rules for working problems on lined paper:

1. Write the part number.
   Then write the letter for each problem.

2. Copy each problem carefully.

3. Box the answer.

4. Leave room between your problems.
   Don't crowd them together.

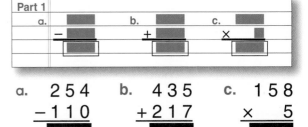

a. $\begin{array}{r} 254 \\ -110 \\ \hline \end{array}$ b. $\begin{array}{r} 435 \\ +217 \\ \hline \end{array}$ c. $\begin{array}{r} 158 \\ \times\ \ \ 5 \\ \hline \end{array}$

Part 2 Write the fraction for each diagram.

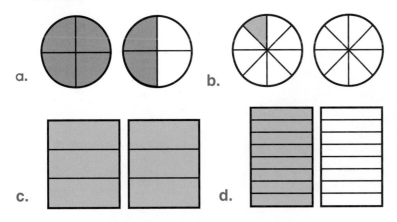

a.

b.

c.

d.

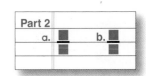

Connecting Math Concepts

Lesson 1

Independent Work

Part 3 Copy and work each problem.

a.   94
   +13

b.   79
    24
   +40

c.   863
  – 47

d.   427
  –352

e.   99
  –56

Part 3	
a.	

Part 4 Write the answer for each fact.

a. 7⟌42   d. 9⟌72   g. 8⟌32   j. 6⟌48   m. 8⟌64

Part 4			
a.		d.	
b.		e.	
c.		f.	

b. 7⟌35   e. 4⟌28   h. 8⟌80   k. 9⟌63   n. 4⟌24

c. 5⟌20   f. 8⟌56   i. 9⟌45   l. 3⟌18   o. 7⟌56

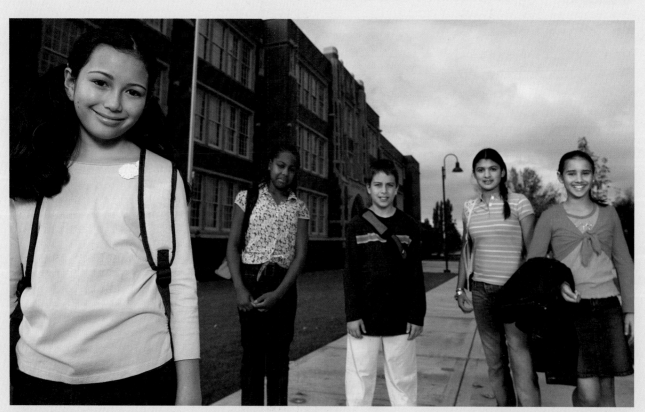

2   Lesson 1   **Connecting Math Concepts**

Lesson 2

Part 1 Write the fraction for each diagram.

a. 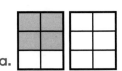 b. 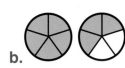 c. 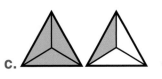 d. 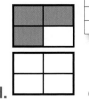 e.

Part 2 Copy each fraction that is more than 1.

a. $\dfrac{4}{5}$ b. $\dfrac{10}{9}$ c. $\dfrac{8}{9}$ d. $\dfrac{7}{6}$ e. $\dfrac{6}{7}$ f. $\dfrac{20}{3}$ g. $\dfrac{12}{4}$

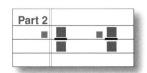

Independent Work

Part 3 Write the fraction for each diagram.

       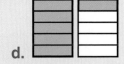

a.   b.   c.   d.

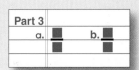

Part 4 Copy and work each problem.

a. $\begin{array}{r} 312 \\ \times \quad 5 \\ \hline \end{array}$ b. $\begin{array}{r} 83 \\ \times \quad 6 \\ \hline \end{array}$ c. $\begin{array}{r} 94 \\ -18 \\ \hline \end{array}$ d. $\begin{array}{r} 126 \\ +979 \\ \hline \end{array}$ e. $\begin{array}{r} 31 \\ \times \quad 9 \\ \hline \end{array}$

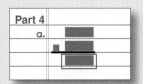

Part 5 Write the answer for each fact.

a. $3\overline{)15}$ d. $9\overline{)54}$ g. $6\overline{)12}$ j. $9\overline{)63}$ m. $7\overline{)49}$

b. $9\overline{)36}$ e. $7\overline{)21}$ h. $8\overline{)64}$ k. $4\overline{)32}$ n. $10\overline{)70}$

c. $8\overline{)16}$ f. $4\overline{)28}$ i. $5\overline{)40}$ l. $6\overline{)48}$ o. $7\overline{)56}$

Lesson

Part 1 Copy and work each problem.

a.	b.	c.	d.	e.
126	219	567	884	653
× 5	× 6	−439	× 7	−584

Part 1	
a.	

Part 2 Write the answer for each fact.

a. 6⟌30 d. 7⟌28 g. 9⟌54 j. 7⟌56 m. 6⟌48

b. 5⟌45 e. 4⟌28 h. 6⟌24 k. 5⟌30 n. 9⟌18

c. 8⟌24 f. 2⟌14 i. 8⟌40 l. 5⟌45 o. 7⟌63

Part 2			
a. ▪		d. ▪	
b. ▪		e. ▪	
c. ▪		f. ▪	

Connecting Math Concepts

Lesson ④

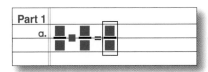

Part 1 Copy the problems you can work. Then work them.

a. $\dfrac{3}{7} + \dfrac{3}{3} = $ ▮

b. $\dfrac{13}{9} - \dfrac{3}{9} = $ ▮

c. $\dfrac{4}{6} + \dfrac{7}{3} = $ ▮

d. $\dfrac{12}{4} - \dfrac{10}{5} = $ ▮

e. $\dfrac{7}{3} + \dfrac{7}{3} = $ ▮

f. $\dfrac{8}{7} + \dfrac{2}{7} = $ ▮

g. $\dfrac{4}{9} - \dfrac{2}{5} = $ ▮

h. $\dfrac{6}{6} - \dfrac{5}{6} = $ ▮

Independent Work

Part 2 Copy and work each problem.

a.
$$158 \times 3$$

b.
$$119 - 76$$

c.
$$485 - 396$$

d.
$$129 \times 4$$

e.
$$210 \\ 14 \\ + 38$$

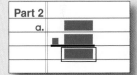

Part 3 Write the answer for each fact.

a. $7\overline{)49}$

d. $6\overline{)36}$

g. $7\overline{)35}$

j. $4\overline{)20}$

m. $6\overline{)54}$

b. $9\overline{)72}$

e. $7\overline{)63}$

h. $10\overline{)60}$

k. $9\overline{)45}$

n. $8\overline{)32}$

c. $7\overline{)21}$

f. $4\overline{)24}$

i. $5\overline{)35}$

l. $8\overline{)56}$

o. $9\overline{)63}$

Lesson 5

Part 1
Write the fraction for each description.

a. The fraction is more than 1. The numbers are 15 and 18.

b. The fraction is less than 1. The numbers are 19 and 20.

c. There are 6 parts shaded. There are 10 parts in each unit.

d. There are 11 parts in each unit. There are 9 parts shaded.

e. The numbers are 14 and 3. The fraction is more than 1.

f. There are 15 parts shaded. There are 14 parts in each unit.

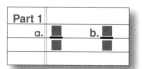

Part 2
Write the place-value equation for each number.

a. 218   b. 112   c. 479   d. 241   e. 111

Part 3
Copy the problems you can work. Then work them.

a. $\dfrac{3}{4} - \dfrac{2}{4} = \blacksquare$ b. $\dfrac{1}{2} + \dfrac{3}{4} = \blacksquare$ c. $\dfrac{7}{3} + \dfrac{1}{3} = \blacksquare$

d. $\dfrac{10}{9} + \dfrac{8}{9} = \blacksquare$ e. $\dfrac{12}{7} - \dfrac{12}{8} = \blacksquare$ f. $\dfrac{15}{2} - \dfrac{4}{2} = \blacksquare$

Independent Work

Part 4
Copy and work each problem.

a.
```
 114
- 96
```
b.
```
  45
  89
+312
```
c.
```
  24
×  2
```
d.
```
  36
×  4
```
e.
```
 736
-281
```

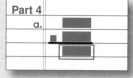

Part 5
Write the answer for each fact.

a. $3\overline{)12}$ d. $4\overline{)28}$ g. $9\overline{)36}$ j. $7\overline{)63}$ m. $4\overline{)28}$

b. $5\overline{)20}$ e. $6\overline{)42}$ h. $7\overline{)63}$ k. $9\overline{)72}$ n. $8\overline{)56}$

c. $6\overline{)36}$ f. $9\overline{)81}$ i. $7\overline{)21}$ l. $6\overline{)24}$ o. $9\overline{)36}$

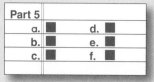

Connecting Math Concepts

Lesson 6

Part 1 Write the fraction for each description.

a. The fraction is less than 1. The numbers are 12 and 14.

b. There are 8 parts in each unit. There are 11 parts shaded.

c. There are 4 parts shaded. There are 7 parts in each unit.

d. The fraction is more than 1. The numbers are 10 and 7.

e. There are 5 parts shaded. There are 9 parts in each unit.

f. The fraction is more than 1. The numbers are 8 and 19.

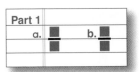

Part 2 Write the place-value equation for each number.

a. 210   b. 70   c. 101   d. 250   e. 700

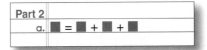

Part 3 Copy the problems you can work. Then work them.

a. $\dfrac{6}{5} - \dfrac{6}{3} = \blacksquare$

b. $\dfrac{18}{7} + \dfrac{2}{7} = \blacksquare$

c. $\dfrac{12}{6} - \dfrac{7}{9} = \blacksquare$

d. $\dfrac{4}{5} - \dfrac{1}{5} = \blacksquare$

e. $\dfrac{4}{11} + \dfrac{7}{11} = \blacksquare$

f. $\dfrac{8}{3} + \dfrac{11}{5} = \blacksquare$

g. $\dfrac{8}{5} + \dfrac{8}{6} = \blacksquare$

h. $\dfrac{16}{9} - \dfrac{16}{9} = \blacksquare$

Part 4 Figure out what the letter equals for the number families you can work.

a. $\xrightarrow{\quad V \qquad P \quad} 367$

b. $\xrightarrow{\quad K \qquad 28 \quad} M$

c. $\xrightarrow{\quad W \qquad 103 \quad} 109$

d. $\xrightarrow{\quad F \qquad 100 \quad} L$

e. $\xrightarrow{\quad 56 \qquad 38 \quad} G$

f. $\xrightarrow{\quad 43 \qquad D \quad} 87$

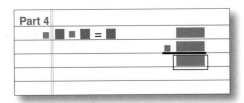

Lesson ⑥

Independent Work

Part 5 Write the answer for each fact.

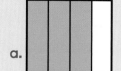   **j.** 10⟌70   **m.** 5⟌25

b. 8⟌56   **e.** 6⟌54   **h.** 6⟌48   **k.** 5⟌30   **n.** 3⟌15

c. 9⟌27   **f.** 8⟌24   **i.** 7⟌56   **l.** 9⟌72   **o.** 8⟌64

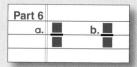

Part 5			
a. ■		d. ■	
b. ■		e. ■	
c. ■		f. ■	

Part 6 Write the fraction for each diagram.

Part 6		
a. ■ ■		b. ■ ■

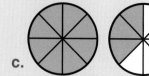

a.

b.

c.

d.

Lesson

Part 1 Copy the problems you can work. Then work them.

a. $\dfrac{35}{8} + \dfrac{4}{5} = \blacksquare$
  b. $\dfrac{11}{8} - \dfrac{9}{10} = \blacksquare$
  c. $\dfrac{4}{5} - \dfrac{3}{5} = \blacksquare$
  d. $\dfrac{15}{3} - \dfrac{10}{8} = \blacksquare$

e. $\dfrac{16}{7} + \dfrac{19}{7} = \blacksquare$
  f. $\dfrac{4}{9} + \dfrac{12}{3} = \blacksquare$
  g. $\dfrac{12}{8} + \dfrac{10}{8} = \blacksquare$
  h. $\dfrac{29}{6} - \dfrac{20}{6} = \blacksquare$

Independent Work

Part 2 Write the place-value equation for each number.

a. 502   b. 40   c. 68   d. 950   e. 75

Part 3 Write each fraction.

a. There are 10 parts in each unit. There are 7 parts shaded.

b. The fraction is less than 1. The numbers are 12 and 9.

c. There are 5 parts shaded. There are 4 parts in each unit.

d. The fraction is more than 1. The numbers are 3 and 20.

e. There are 8 parts shaded. There are 9 parts in each unit.

f. The fraction is more than 1. The numbers are 11 and 6.

Lesson

Part 1   Write a division problem for each fraction and work it.

a. $\dfrac{14}{7}$ b. $\dfrac{40}{5}$ c. $\dfrac{48}{8}$ d. $\dfrac{32}{4}$ e. $\dfrac{21}{3}$

Independent Work

Part 2   Write the place-value equation for each number.

a. 36   b. 111   c. 240   d. 583   e. 301

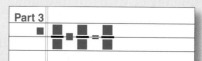

Part 3   Copy the problems you can work. Then work them.

a. $\dfrac{3}{8} + \dfrac{4}{8} = \blacksquare$ b. $\dfrac{10}{9} - \dfrac{2}{9} = \blacksquare$ c. $\dfrac{5}{7} + \dfrac{8}{3} = \blacksquare$

d. $\dfrac{5}{10} + \dfrac{8}{10} = \blacksquare$ e. $\dfrac{11}{6} - \dfrac{2}{3} = \blacksquare$ f. $\dfrac{7}{3} + \dfrac{7}{3} = \blacksquare$

Connecting Math Concepts

Lesson 9

Part 1 Write each fraction.

a. The fraction equals 3. The bottom number is 20.

b. The bottom number is 10. The fraction equals 11.

c. The fraction equals 6. The bottom number is 100.

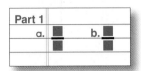

Part 1		
a. ▮▮	b. ▮▮

Part 2 Figure out what each letter equals.

a. $\dfrac{T \quad\quad 41}{} \rightarrow 83$

b. $\dfrac{52 \quad\quad 18}{} \rightarrow D$

c. $\dfrac{29 \quad\quad R}{} \rightarrow 173$

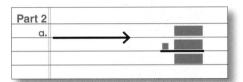

Part 3 Write a division problem for each fraction and work it.

a. $\dfrac{18}{3}$ b. $\dfrac{28}{4}$ c. $\dfrac{35}{5}$ d. $\dfrac{21}{7}$ e. $\dfrac{48}{6}$

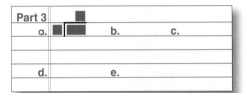

Part 4 Make the number family. Figure out what the letter equals.

a. The big number is M. The small numbers are 52 and 39. What does M equal?

b. The small numbers are Y and 13. The big number is 100. What does Y equal?

c. The big number is 97. The small numbers are 15 and D. What does D equal?

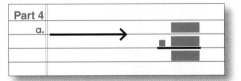

Part 5 Copy and work each problem.

a. $\dfrac{4}{2} \times \dfrac{3}{9} = \dfrac{\blacksquare}{\blacksquare}$

b. $\dfrac{2}{8} \times \dfrac{7}{5} = \dfrac{\blacksquare}{\blacksquare}$

c. $\dfrac{2}{3} \times \dfrac{4}{5} = \dfrac{\blacksquare}{\blacksquare}$

d. $\dfrac{1}{3} \times \dfrac{5}{1} = \dfrac{\blacksquare}{\blacksquare}$

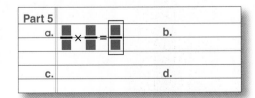

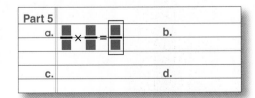

Connecting Math Concepts

Lesson

Part 6 Write the place-value equation for each number.

a. 320   b. 159   c. 201   d. 26   e. 410

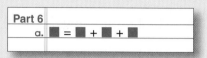

Part 7 Copy the problems you can work. Then work them.

a. $\dfrac{16}{11} - \dfrac{5}{8} = $ ▉

b. $\dfrac{10}{2} + \dfrac{7}{2} = $ ▉

c. $\dfrac{10}{4} - \dfrac{5}{6} = $ ▉

d. $\dfrac{9}{8} + \dfrac{7}{8} = $ ▉

e. $\dfrac{9}{10} + \dfrac{6}{10} = $ ▉

f. $\dfrac{14}{8} - \dfrac{14}{9} = $ ▉

Part 8 Write each fraction.

a. There are 6 parts shaded. There are 9 parts in each unit.

b. The fraction is more than 1. The numbers are 15 and 10.

c. The fraction is less than 1. The numbers are 16 and 14.

d. There are 9 parts in each unit. There are 12 parts shaded.

e. The fraction is more than 1. The numbers are 17 and 19.

f. There are 4 parts shaded. There are 11 parts in each unit.

Lesson 10

Part 1 Write each fraction.

a. The fraction equals 11. The bottom number is 7.

b. The fraction equals 8. The bottom number is 4.

c. The bottom number is 9. The fraction equals 7.

d. The fraction equals 10. The bottom number is 3.

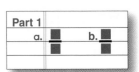

Part 2 Make a number family. Figure out the missing number.

a. The big number is 87. The small numbers are K and N.

$$\boxed{K = 49}$$

b. The small numbers are V and 106. The big number is P.

$$\boxed{P = 238}$$

c. 56 and Q are small numbers. W is the big number.

$$\boxed{Q = 80}$$

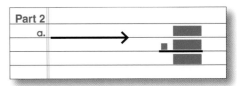

Part 3 Copy and work each problem.

a. $\dfrac{2}{3} \times \dfrac{7}{5} = \dfrac{\blacksquare}{\blacksquare}$

d. $\dfrac{2}{1} \times \dfrac{4}{7} = \dfrac{\blacksquare}{\blacksquare}$

b. $\dfrac{1}{4} \times \dfrac{6}{3} = \dfrac{\blacksquare}{\blacksquare}$

e. $\dfrac{7}{3} \times \dfrac{4}{9} = \dfrac{\blacksquare}{\blacksquare}$

c. $\dfrac{10}{3} \times \dfrac{8}{6} = \dfrac{\blacksquare}{\blacksquare}$

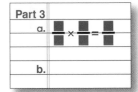

Part 4 Make a number family with two letters and a number.

a. Ann had 4 fewer cards than Bob had.

b. Hal had 23 fewer cards than Pam had.

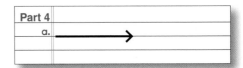

Lesson 10

Part 5
Write the place-value equation for each number.

a. 804   b. 650   c. 39   d. 12   e. 81

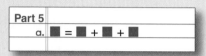

Part 6
Copy the problems you can work. Then work them.

a. $\frac{10}{3} + \frac{9}{3} = \blacksquare$ b. $\frac{16}{5} - \frac{8}{5} = \blacksquare$ c. $\frac{1}{2} + \frac{1}{3} = \blacksquare$

d. $\frac{6}{10} - \frac{5}{8} = \blacksquare$ e. $\frac{7}{8} + \frac{6}{8} = \blacksquare$ f. $\frac{8}{9} + \frac{4}{9} = \blacksquare$

Part 7
Copy and work each problem.

a.
$$\begin{array}{r} 417 \\ \times\quad 3 \\ \hline \end{array}$$

b.
$$\begin{array}{r} 549 \\ +\quad 43 \\ \hline \end{array}$$

c.
$$\begin{array}{r} 810 \\ -\quad 50 \\ \hline \end{array}$$

d.
$$\begin{array}{r} 119 \\ \times\quad 5 \\ \hline \end{array}$$

Part 8
Write the fraction for each description.

a. The fraction is less than 1. The numbers are 16 and 12.

b. The numbers are 10 and 9. The fraction is more than 1.

c. There are 6 parts in each unit. 11 parts are shaded.

d. There are 20 parts shaded. There are 10 parts in each unit.

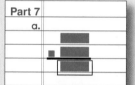

Lesson 11

Part 1 Make a number family with two letters and a number.

a. Bob had 2 fewer quarters than Nan had.

b. Vern had 36 more dollars than Jill had.

c. Jack is 13 inches shorter than Tom.

d. The church is 41 feet taller than the bank.

e. Tim was 4 years older than Pam.

f. Jill is 31 years younger than Kay.

Part 1	
a.	⟶

Part 2 Copy and work each problem.

a. $\dfrac{9}{2} \times \dfrac{5}{4} = \dfrac{\blacksquare}{\blacksquare}$

d. $\dfrac{4}{5} \times \dfrac{4}{1} = \dfrac{\blacksquare}{\blacksquare}$

b. $\dfrac{1}{5} \times \dfrac{3}{8} = \dfrac{\blacksquare}{\blacksquare}$

e. $\dfrac{2}{2} \times \dfrac{3}{7} = \dfrac{\blacksquare}{\blacksquare}$

c. $\dfrac{10}{3} \times \dfrac{8}{9} = \dfrac{\blacksquare}{\blacksquare}$

Part 2	
a.	■ × ■ = ■

Part 3 Make a number family. Figure out the missing number.

a. M and G are the small numbers. 92 is the big number.

M = 34

b. The small numbers are 148 and P. F is the big number.

P = 78

c. The big number is E. T and 75 are the small numbers.

E = 199

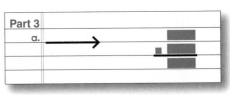

Lesson 12

Part 1
Make a number family with two letters and a number.

a. Pile A weighs 34 pounds more than pile B.

b. Tree A was 23 feet shorter than tree B.

c. The worm was 13 inches shorter than the snake.

d. Street A is 7 feet wider than street B.

Part 2
Copy the problems you can work. Then work them.

a. $\dfrac{2}{5} \times \dfrac{3}{8} = \blacksquare$

b. $\dfrac{13}{6} - \dfrac{10}{6} = \blacksquare$

c. $\dfrac{9}{4} - \dfrac{6}{4} = \blacksquare$

d. $\dfrac{7}{11} + \dfrac{4}{11} = \blacksquare$

e. $\dfrac{3}{8} + \dfrac{4}{8} = \blacksquare$

f. $\dfrac{5}{6} \times \dfrac{2}{6} = \blacksquare$

g. $\dfrac{3}{5} + \dfrac{3}{7} = \blacksquare$

h. $\dfrac{1}{7} + \dfrac{12}{7} = \blacksquare$

Part 3
Make a number family. Figure out the missing number.

a. The small numbers are K and 64. The big number is P.

P = 129

b. The big number is M. The small numbers are 16 and R.

R = 48

c. The small numbers are J and P. The big number is 211.

J = 88

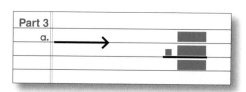

Independent Work

Part 4
Write each fraction.

a. The fraction equals 8. The bottom number is 5.

b. The fraction is less than 1. The numbers are 5 and 10.

c. The bottom number of the fraction is 3. The fraction equals 9.

d. There are 7 parts in each unit. 9 parts are shaded.

e. The fraction equals 6. The bottom number is 3.

Connecting Math Concepts

Lesson

Copy the problems you can work. Then work those problems.

a. $\dfrac{4}{5} + \dfrac{7}{5} = \blacksquare$ b. $\dfrac{7}{8} - \dfrac{5}{9} = \blacksquare$ c. $\dfrac{2}{3} \times \dfrac{9}{3} = \blacksquare$ d. $\dfrac{8}{3} - \dfrac{4}{3} = \blacksquare$

e. $\dfrac{2}{5} + \dfrac{7}{8} = \blacksquare$ f. $\dfrac{9}{7} - \dfrac{5}{7} = \blacksquare$ g. $\dfrac{5}{6} + \dfrac{8}{6} = \blacksquare$ h. $\dfrac{3}{9} \times \dfrac{6}{1} = \blacksquare$

Part 2 Make a number family. Figure out the number that answers the question.

a. Richard has 95 dollars less than Susan. Richard has 140 dollars. How many dollars does Susan have?

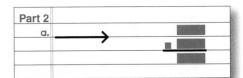

b. Bob weighs 14 pounds less than Mike. Mike weighs 124 pounds. How many pounds does Bob weigh?

c. The truck went 23 miles farther than the car went. The car went 90 miles. How many miles did the truck go?

Part 3 Figure out what each fraction equals.

a. $\dfrac{186}{6} = \blacksquare$ b. $\dfrac{428}{4} = \blacksquare$ c. $\dfrac{99}{3} = \blacksquare$

Independent Work

Part 4 Write each fraction.

a. The fraction equals 10. The bottom number is 7.

b. There are 2 parts in each unit. 3 parts are shaded.

c. The bottom number of the fraction is 3. The fraction equals 12.

d. The fraction is more than 1. The numbers are 12 and 11.

e. The bottom number of the fraction is 8. The fraction equals 2.

Lesson 13

Part 5   Make a number family. Figure out the missing number.

a. The big number is 239.
The small numbers are B and C.

   | C = 118 |

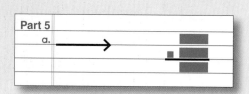

b. The small numbers are 71 and P.
The big number is B.

   | B = 302 |

c. The big number is P.
The small numbers are 114 and T.

   | P = 289 |

Part 6   Write the fraction for each picture.

a.

b.

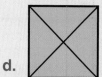

c.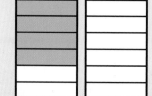

d.

Lesson 14

Part 1   Make a number family. Figure out the number that answers the question.

a. Scott Lake is 13 miles long. Blue Lake is 24 miles longer than Scott Lake. How many miles long is Blue Lake?

b. Sid has 95 baseball cards. Peter had 68 fewer baseball cards than Sid. How many cards does Peter have?

c. Ted walked 14 miles farther than Maria. Ted walked 25 miles. How many miles did Maria walk?

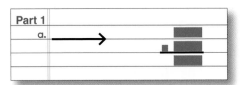

Part 2   Copy the problems you can work. Then work them.

a. $\dfrac{9}{8} - \dfrac{4}{5} = \blacksquare$

b. $\dfrac{3}{4} + \dfrac{2}{4} = \blacksquare$

c. $\dfrac{9}{3} \times \dfrac{2}{6} = \blacksquare$

d. $\dfrac{8}{5} - \dfrac{7}{5} = \blacksquare$

e. $\dfrac{1}{2} \times \dfrac{3}{2} = \blacksquare$

f. $\dfrac{5}{9} + \dfrac{7}{7} = \blacksquare$

g. $\dfrac{2}{6} + \dfrac{5}{6} = \blacksquare$

h. $\dfrac{3}{5} - \dfrac{1}{2} = \blacksquare$

Independent Work

Part 3   Copy and complete each equation.

a. $\dfrac{316}{4} = \blacksquare$

b. $\dfrac{77}{7} = \blacksquare$

c. $\dfrac{100}{5} = \blacksquare$

d. $\dfrac{86}{2} = \blacksquare$

Part 4   Write each fraction.

a. The fraction equals 5. The bottom number is 3.

b. There are 2 parts in each unit. 2 parts are shaded.

c. The bottom number of the fraction is 1. The fraction equals 24.

d. The fraction is less than 1. The numbers are 20 and 19.

e. The fraction equals 2. The bottom number is 20.

Lesson 14

Part 5 Make a number family. Figure out the missing number.

a. The big number is K.
The small numbers are 103 and M.

$$\boxed{K = 219}$$

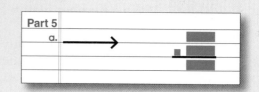

b. The small numbers are P and 29.
The big number is N.

$$\boxed{P = 201}$$

c. The big number is 98.
The small numbers are R and J.

$$\boxed{R = 29}$$

Part 6 Copy and work each problem.

a. $4\overline{)840}$ b. $3\overline{)906}$ c. $4\overline{)168}$

d. $7\overline{)280}$ e. $8\overline{)648}$

Connecting Math Concepts

Lesson 15

Part 1 Make a number family. Figure out the number that answers the question.

a. Roberto made 12 more cookies than Sammy made. Roberto made 36 cookies. How many cookies did Sammy make?

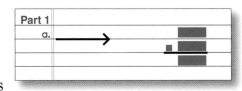

b. Isabel earned 200 dollars. Timmy earned 40 fewer dollars than Isabel earned. How many dollars did Timmy earn?

c. Cristi sold 89 plants. Cristi sold 53 fewer plants than Mandy sold. How many plants did Mandy sell?

Independent Work

Part 2 Write each fraction.

a. The fraction is less than one. The numbers are 37 and 38.

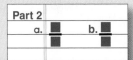

b. There are 7 parts in each unit. 12 parts are shaded.

c. The fraction equals 3. The numbers are 9 and 27.

d. The bottom number of the fraction is 48. The fraction equals 2.

Part 3 Copy the problems you can work. Then work them.

a. $\dfrac{2}{3} \times \dfrac{7}{8} = $ ■
b. $\dfrac{17}{5} - \dfrac{1}{5} = $ ■
c. $\dfrac{9}{9} + \dfrac{7}{8} = $ ■

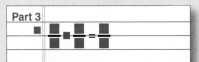

d. $\dfrac{14}{2} - \dfrac{5}{9} = $ ■
e. $\dfrac{17}{4} - \dfrac{10}{4} = $ ■
f. $\dfrac{3}{4} \times \dfrac{4}{3} = $ ■

Part 4 Make a number family. Figure out the missing number.

a. The big number is K. The small numbers are Z and 136.

$Z = 402$

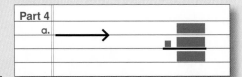

b. The small numbers are C and 111. The big number is G.

$G = 202$

Part 5 Copy and complete each equation.

a. $\dfrac{144}{6} = \blacksquare$

b. $\dfrac{77}{7} = \blacksquare$

c. $\dfrac{256}{8} = \blacksquare$

Part 6 Make a number family with two letters and a number.

a. Ricardo was 40 months younger than Henry.

b. Taylor earned $898 more than Mark earned.

c. The red boat was 17 feet longer than the black boat.

d. Sidney had 34 fewer stamps than Jorge had.

e. Ted's house was 46 feet narrower than Chi's house.

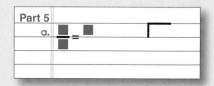

Connecting Math Concepts

Lesson

Part 1 Make a number family. Figure out the missing number.

a. The big number is 406. The small numbers are P and Q.

$$Q = 14$$

b. The small numbers are 84 and R. The big number is T.

$$T = 156$$

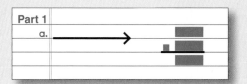

Part 2 Write the answer for each fact.

a. $8\overline{)3\,2}$ b. $6\overline{)3\,6}$ c. $5\overline{)3\,5}$ d. $9\overline{)8\,1}$

e. $7\overline{)4\,2}$ f. $9\overline{)6\,3}$ g. $5\overline{)5\,0}$ h. $7\overline{)5\,6}$

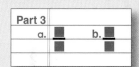

Part 3 Write each fraction.

a. The fraction equals 1. The bottom number is 10.

b. There are 7 parts in each unit. 5 parts are shaded.

c. The fraction equals 6. The bottom number is 3.

Lesson

Part 1 Work each problem. Write the answer as a number and a unit name.

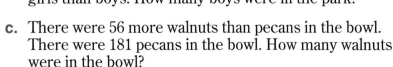

a. The tree was 88 feet tall. The tree was 42 feet shorter than the building. How tall was the building?

b. There were 78 girls in the park. There were 22 more girls than boys. How many boys were in the park?

c. There were 56 more walnuts than pecans in the bowl. There were 181 pecans in the bowl. How many walnuts were in the bowl?

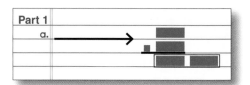

Part 2 Copy and work each problem.

a.
$$\begin{array}{r} 2\,3 \\ \times\ 8\,0 \\ \hline \end{array}$$

b.
$$\begin{array}{r} 4\,5 \\ \times\ 3\,0 \\ \hline \end{array}$$

c.
$$\begin{array}{r} 9\,1 \\ \times\ 7\,0 \\ \hline \end{array}$$

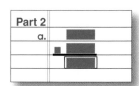

Independent Work

Part 3 Write each fraction.

a. The denominator of the fraction is 10. The fraction equals 2.

b. The fraction equals 8. The denominator is 6.

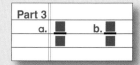

Part 4 Make a number family. Figure out the missing number.

a. The big number is F. The small numbers are R and 158.

   $\boxed{F = 216}$

b. The small numbers are T and B. The big number is 96.

   $\boxed{T = 21}$

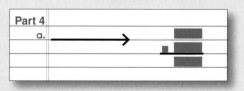

Part 5 Copy and complete each equation.

a. $\dfrac{\blacksquare}{4} = 3$

b. $\dfrac{\blacksquare}{7} = 1$

c. $\dfrac{\blacksquare}{9} = 5$

d. $\dfrac{\blacksquare}{7} = 7$

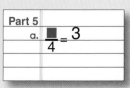

Connecting Math Concepts

Lesson 18

Part 1 Work each problem. Write the answer as a number and a unit name.

a. The tortoise was 146 years old. The horse was 28 years old. How much older was the tortoise than the horse?

b. The dog weighed 42 pounds. The rat weighed 3 pounds. How much heavier was the dog than the rat?

c. Maria is 64 years old. Jan is 11 years old. How much younger is Jan than Maria?

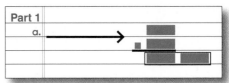

Part 2 Write the number that answers each question.

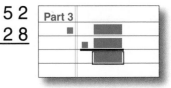

	January	February	Total for both months
Hill Park	35	14	49
River Park	15	18	33
Total for both parks	50	32	82

Number of People Visiting Two Parks

a. What's the number of people who visited River Park in February?

b. How many people visited both parks in January?

c. How many people visited Hill Park in January?

d. What's the total number of people who visited River Park?

e. What's the total number of visitors for both parks in February?

Part 3 Copy and work the problems that multiply by a tens number ending in zero.

a.
```
    2 3
  × 6 0
```

b.
```
  6 7 3
  ×   9
```

c.
```
    4 7
  × 3 0
```

d.
```
    5 6
  × 9 0
```

e.
```
    5 2
  × 2 8
```

Lesson 18

Part 4 Write the answer for each fact.

a. $8\overline{)24}$ b. $4\overline{)36}$ c. $6\overline{)24}$ d. $9\overline{)54}$

e. $7\overline{)28}$ f. $7\overline{)63}$ g. $8\overline{)72}$ h. $8\overline{)64}$

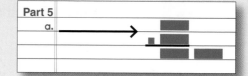

Part 5 Make a number family. Work the problem. Remember the unit name.

a. Mr. Briggs weighed 48 pounds more than Mrs. Jones. Mrs. Jones weighed 134 pounds. How much did Mr. Briggs weigh?

b. The red boat was 56 feet long. The green boat was 21 feet shorter than the red boat. How many feet long was the green boat?

c. In January Jay ran 93 miles less than he ran in February. In January Jay ran 175 miles. How many miles did he run in February?

Part 6 Write each fraction.

a. The fraction equals 7. The denominator is 3.

b. The denominator of the fraction is 5. The fraction equals 5.

Part 7 Copy and complete each equation.

a. $\dfrac{\blacksquare}{5} = 2$ b. $\dfrac{\blacksquare}{10} = 1$ c. $\dfrac{\blacksquare}{9} = 8$ d. $\dfrac{\blacksquare}{2} = 20$

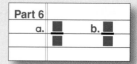

Lesson 19

Part 1 Work each problem. Write the answer as a number and a unit name.

a. The bracelet is 136 years old. The ring is 98 years old. How much older is the bracelet than the ring?

b. Tim worked 56 hours. Heidi worked 23 more hours than Tim worked. How many hours did Heidi work?

c. The trip to the mountains was 127 miles. The trip to the beach was 39 miles. How much farther was it to the mountains than the beach?

d. They saw 36 more deer than elk. They saw 62 elk. How many deer did they see?

Part 2 Write the number that answers each question.

	Boys	Girls	Total children
Allen Bridge	42	97	139
Toll Bridge	68	43	111
Total for both bridges	110	140	250

Number of Children Walking Across Two Bridges

a. How many girls walked across Toll Bridge?

b. How many children walked across Allen Bridge?

c. How many boys walked across Allen Bridge?

d. How many girls walked across both bridges?

e. How many children walked across both bridges?

Part 2		
a.		b.

Independent Work

Part 3 Copy and complete each equation.

a. $\dfrac{\blacksquare}{9} = 6$ b. $\dfrac{\blacksquare}{4} = 8$ c. $\dfrac{\blacksquare}{12} = 20$ d. $\dfrac{\blacksquare}{36} = 8$

Part 3	
a.	$\dfrac{\blacksquare}{9} = 6$

Lesson 19

Part 4
Write the answer for each fact.

a. $4\overline{)32}$ b. $4\overline{)40}$ c. $8\overline{)40}$ d. $9\overline{)81}$

e. $6\overline{)36}$ f. $9\overline{)90}$ g. $9\overline{)36}$ h. $7\overline{)56}$

Part 4		
	a.	b.

Part 5
Write each fraction.

a. The fraction equals 3. The denominator is 10.

b. The denominator of the fraction is 8. The fraction equals 4.

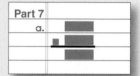

Part 6
Write the place value equation for each number.

a. 101   b. 593   c. 110

Part 7
Copy and work each problem.

a. $\begin{array}{r} 79 \\ \times\ 80 \\ \hline \end{array}$ b. $\begin{array}{r} 620 \\ \times\ \ 46 \\ \hline \end{array}$ c. $\begin{array}{r} 81 \\ \times\ 36 \\ \hline \end{array}$

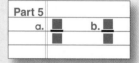

Lesson 20

Part 1 Copy and work each problem.

a. $7\overline{)99}$ b. $8\overline{)252}$ c. $5\overline{)423}$ d. $6\overline{)382}$

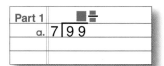

Independent Work

Part 2 Make a number family. Work the problem. Remember the unit name.

a. Jan's shoes cost $56 more than her blouse. Her blouse cost $27. How much did her shoes cost?

b. The spelling book had 88 pages. The math book had 268 pages. How many more pages did the math book have?

c. The gator had 80 teeth. The mouse had 16 teeth. How many fewer teeth did the mouse have?

d. There were 186 squirrels in the field. There were 113 fewer squirrels in the field than in the woods. How many squirrels were in the woods?

Part 3 Write each fraction.

a. The fraction equals 11. The denominator is 3.

b. The denominator of the fraction is 9. The fraction equals 10.

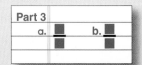

Lesson 21

Part 1
Write two equations for each point.

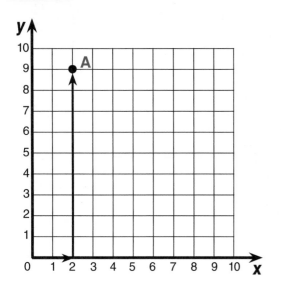

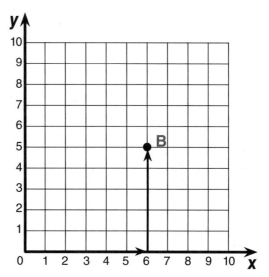

Part 1	
A.	$x = \blacksquare$, $y = \blacksquare$
B.	$x = \blacksquare$, $y = \blacksquare$

Part 2
Write each fraction as a division problem. Figure out the mixed number it equals.

a. $\dfrac{79}{9}$ b. $\dfrac{214}{7}$ c. $\dfrac{69}{8}$ d. $\dfrac{37}{6}$

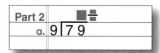

Independent Work

Part 3
Write the answer to each division fact.

a. $7\overline{)6\ 3}$ b. $9\overline{)2\ 7}$ c. $9\overline{)8\ 1}$ d. $8\overline{)4\ 0}$

e. $7\overline{)5\ 6}$ f. $6\overline{)5\ 4}$ g. $9\overline{)3\ 6}$ h. $8\overline{)6\ 4}$

Part 3	
a.	b.

Lesson 21

Part 4 Copy and work each problem.

a. $\begin{array}{r} 70 \\ \times\ 38 \\ \hline \end{array}$

b. $\begin{array}{r} 26 \\ \times\ 13 \\ \hline \end{array}$

c. $\begin{array}{r} 182 \\ \times\ 22 \\ \hline \end{array}$

d. $\begin{array}{r} 98 \\ \times\ 50 \\ \hline \end{array}$

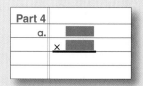

Part 5 Make a number family. Work the problem. Remember the unit name.

a. The desk cost $194. The desk cost $138 more than the bench. How much did the bench cost?

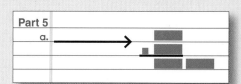

b. The car went 275 miles. The bus went 154 miles. How many more miles did the car go?

c. Myra earned $56 less than Elise. Myra earned $307. How much did Elise earn?

Part 6 Copy the problems you can work. Then work them.

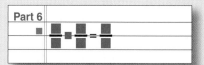

a. $\dfrac{3}{12} \times \dfrac{5}{3} = $ ▮

b. $\dfrac{26}{5} - \dfrac{16}{5} = $ ▮

c. $\dfrac{48}{3} + \dfrac{18}{5} = $ ▮

d. $\dfrac{40}{9} - \dfrac{17}{9} = $ ▮

Lesson

Part 1 Work each problem.

a. There are 4 geese for every 5 ducks. If there were 30 ducks, how many geese were there?

b. Every 2 shirts have 8 buttons. How many buttons are on 12 shirts?

c. Every 3 boxes weigh 10 pounds. How many pounds do 9 boxes weigh?

Part 2 Write the x and y equation for each point.

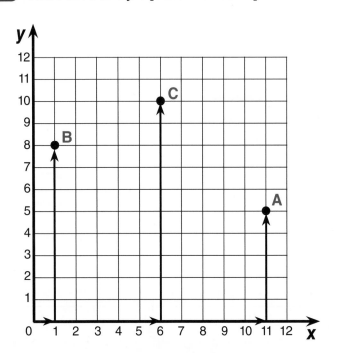

Part 3 Write each fraction as a division problem. Figure out the mixed number it equals.

a. $\dfrac{389}{9}$ b. $\dfrac{97}{3}$ c. $\dfrac{78}{5}$ d. $\dfrac{283}{4}$

Lesson 22

Independent Work

Part 4 Copy and work each problem. Rewrite each whole number as a fraction.

a.
$$4$$
$$-\frac{3}{5}$$

b.
$$\frac{5}{7}$$
$$+\ 3$$

c.
$$9$$
$$+\frac{7}{5}$$

d.
$$\frac{45}{5}$$
$$-\ 3$$

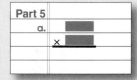

Part 5 Copy and work each problem.

a.
$$38$$
$$\times\ 14$$

b.
$$176$$
$$\times\ 40$$

c.
$$89$$
$$\times\ 39$$

d.
$$70$$
$$\times\ 50$$

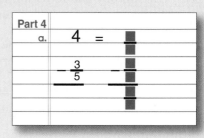

Part 6 Write the answer to each division fact.

a. $9\overline{)54}$ b. $9\overline{)72}$ c. $6\overline{)42}$ d. $4\overline{)36}$

Part 6	
a.	b.

e. $7\overline{)56}$ f. $7\overline{)21}$ g. $6\overline{)48}$ h. $8\overline{)64}$

Part 7 Answer each question.

Part 7	
a.	b.

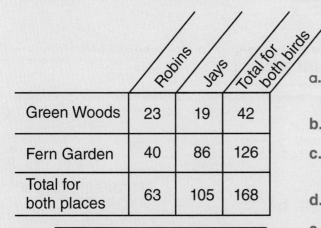

	Robins	Jays	Total for both birds
Green Woods	23	19	42
Fern Garden	40	86	126
Total for both places	63	105	168

Number of Birds Seen in Two Places

a. What is the total number of robins seen in both places?

b. How many jays were seen in Fern Garden?

c. What was the total number of birds seen in Green Woods?

d. How many robins were seen in Fern Garden?

e. How many birds were seen in both places?

Lesson 23

Part 1 Work each problem.

a. The factory made 9 clocks every 2 minutes. How many clocks would the factory make in 6 minutes?

b. There were 5 fleas for every 3 ants. If there were 27 ants, how many fleas were there?

c. Every 7 days the restaurant uses 4 melons. How many melons does the restaurant use in 35 days?

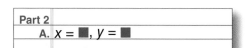

Part 2 Write an *x* equation and *y* equation for each point.

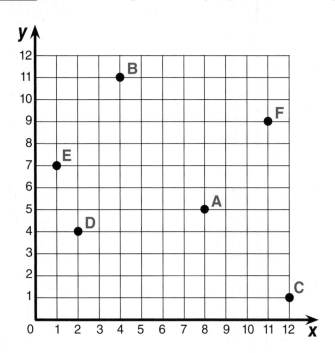

Part 2	
A.	$x = \blacksquare,\ y = \blacksquare$

Independent Work

Part 3 Write the answer to each division fact.

a. $6\overline{)36}$ b. $7\overline{)56}$ c. $7\overline{)63}$ d. $6\overline{)18}$

e. $8\overline{)48}$ f. $7\overline{)49}$ g. $7\overline{)35}$ h. $6\overline{)42}$

Part 3	
a.	b.

Part 4 Write each fraction as a division problem. Figure out the mixed number it equals.

a. $\dfrac{16}{5}$ b. $\dfrac{30}{9}$ c. $\dfrac{56}{9}$ d. $\dfrac{49}{6}$ e. $\dfrac{54}{5}$

Part 5 Work each problem. Remember the unit name.

a. Jon had $356. Sid had $83 more than Jon. How much money did Sid have?

b. The train car weighed 186 tons. The bus weighed 63 tons. How much heavier was the train car than the bus?

c. Jane weighed 144 pounds less than Louis. Jane weighed 129 pounds. How much did Louis weigh?

Part 6 Copy and work each problem.

a. $\dfrac{41}{5} + \dfrac{12}{5} = $ ▮   b. $\dfrac{24}{3} \times \dfrac{1}{3} = $ ▮   c. $\dfrac{12}{17} - \dfrac{4}{17} = $ ▮   d. $\dfrac{5}{8} \times \dfrac{7}{11} = $ ▮

Part 7 Copy and work each problem.

a. $\begin{array}{r} 12 \\ \times\ 18 \\ \hline \end{array}$ b. $\begin{array}{r} 24 \\ \times\ 36 \\ \hline \end{array}$ c. $\begin{array}{r} 506 \\ \times\ \ 30 \\ \hline \end{array}$ d. $\begin{array}{r} 87 \\ \times\ 42 \\ \hline \end{array}$ e. $\begin{array}{r} 349 \\ \times\ \ 12 \\ \hline \end{array}$

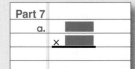

Lesson 24

Part 1 — Copy and work each problem.

a.
$$\dfrac{6}{4}$$
$$+\dfrac{2}{3}$$

b.
$$\dfrac{7}{2}$$
$$-\dfrac{3}{5}$$

c.
$$\dfrac{3}{2}$$
$$+\dfrac{5}{4}$$

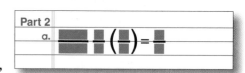

Part 2 — Work each problem.

a. The ratio of cups to plates is 2 to 3. If there are 16 cups, how many plates are there?

b. The ratio of lions to tigers is 7 to 5. If there are 45 tigers, how many lions are there?

c. The ratio of skunks to spiders is 8 to 9. How many skunks are there if there are 45 spiders?

Part 3 — Answer each question.

	Fifth graders	Sixth graders	Total for both grades
Jefferson School	128	223	351
Madison School	155	184	339
Total for both schools	283	407	690

Fifth Graders and Sixth Graders at Two Schools

a. In which school are there more fifth graders?

b. What's the total number of sixth graders for both schools?

c. How many sixth graders attend Jefferson school?

d. What's the total number of fifth and sixth graders who attend Madison school?

e. In which school are there fewer sixth graders?

Lesson 24

Independent Work

Part 4 Write each fraction as a division problem. Figure out the mixed number it equals.

a. $\dfrac{88}{9} =$ b. $\dfrac{45}{6} =$ c. $\dfrac{75}{8} =$ d. $\dfrac{20}{6} =$ e. $\dfrac{86}{9} =$

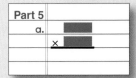

Part 5 Copy and work each problem.

a. $\begin{array}{r} 64 \\ \times\ 70 \\ \hline \end{array}$ b. $\begin{array}{r} 48 \\ \times\ 83 \\ \hline \end{array}$ c. $\begin{array}{r} 27 \\ \times\ 56 \\ \hline \end{array}$ d. $\begin{array}{r} 431 \\ \times\ \ \ 82 \\ \hline \end{array}$

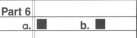

Part 6 Write the answer to each division fact.

a. $7\overline{)63}$ b. $8\overline{)48}$ c. $8\overline{)64}$ d. $5\overline{)25}$

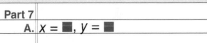

e. $8\overline{)40}$ f. $9\overline{)72}$ g. $7\overline{)35}$ h. $9\overline{)81}$

Part 7 Write the x and y equation for each point.

Part 7

A. $x = \blacksquare,\ y = \blacksquare$

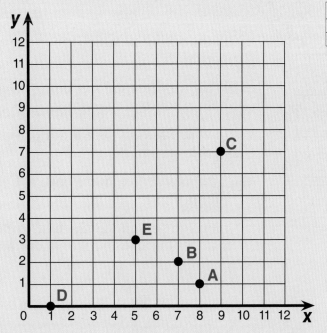

Lesson 25

Part 1 Copy and work each problem.

a. $\dfrac{5}{6}$
$-\dfrac{3}{8}$

b. $\dfrac{4}{3}$
$+\dfrac{1}{9}$

c. $\dfrac{8}{10}$
$-\dfrac{3}{4}$

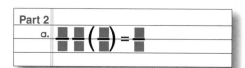

Part 2 Work each problem. Use initials for the names.

a. The ratio of boys to men is 6 to 3. If there are 15 men, how many boys are there?

b. The ratio of sheep to farmers is 9 to 2. How many sheep are there for 10 farmers?

c. There are 5 knives for every 11 spoons. If there are 15 knives, how many spoons are there?

Independent Work

Part 3 Answer each question.

	Saddle Ranch	Camp Ranch	Total for both ranches
Winter	65	30	95
Spring	26	41	67
Total for both seasons	91	71	162

Calves Born at Two Ranches

a. In the spring, at which ranch were more calves born?

b. What was the total number of calves born at both ranches in the winter?

c. In which season were fewer calves born at Saddle Ranch?

d. How many calves were born at Camp Ranch in the spring?

e. In which season was the total greater for the number of calves born at both ranches?

Part 4 Copy and work each problem.

a. $\dfrac{71}{12} + \dfrac{8}{12} = $ ▮

b. $\dfrac{26}{32} - \dfrac{5}{32} = $ ▮

c. $\dfrac{24}{8} \times \dfrac{1}{7} = $ ▮

d. $\dfrac{15}{7} \times \dfrac{10}{7} = $ ▮

Part 5 Work each problem. Remember the unit name.

a. The snake was 48 inches long. The snake was 23 inches shorter than the rake. How long was the rake?

b. At the end of the day, there were 497 fewer gallons of water in the tank than in the morning. There were 876 gallons in the tank in the morning. How many gallons were in the tank at the end of the day?

c. There were 420 trees in grove A. Grove B had 116 trees. How many more trees were in grove A than in grove B?

Part 6 Write the answer to each division fact.

a. $3\overline{)24}$ b. $8\overline{)72}$ c. $6\overline{)54}$ d. $6\overline{)36}$

e. $3\overline{)30}$ f. $7\overline{)42}$ g. $8\overline{)40}$ h. $3\overline{)15}$

Lesson 25

Part 7 Write the *x* and *y* equation for each point.

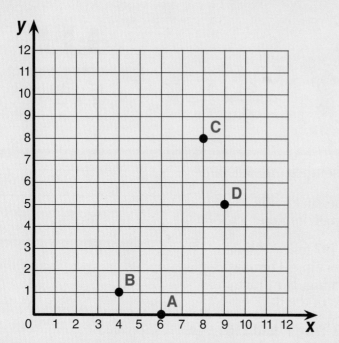

Part 7	
A.	$x = \blacksquare$, $y = \blacksquare$

Lesson 26

Part 1 Copy and work each problem.

a. $\dfrac{5}{2}$

$-\dfrac{2}{7}$

b. $\dfrac{7}{4}$

$+\dfrac{5}{6}$

c. $\dfrac{11}{12}$

$+\dfrac{1}{4}$

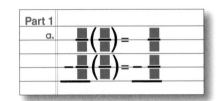

Part 2 Copy each number. Then write the closest tens number.

a. 348   b. 24   c. 107   d. 252   e. 6   f. 198

Part 3 Work each problem. Use initials for the names.

a. There are 12 bears for every 5 dogs. How many bears are there for 20 dogs?

b. Every 2 books weigh 7 pounds. How many pounds do 18 books weigh?

c. The ratio of buds to flowers is 1 to 8. If there are 48 flowers, how many buds are there?

d. 10 pens cost 4 dollars. How many pens cost 28 dollars?

Part 4 Write each mixed number as a decimal value.

a. $7\dfrac{14}{1000}$

b. $19\dfrac{13}{100}$

c. $23\dfrac{6}{10}$

d. $60\dfrac{8}{100}$

e. $15\dfrac{400}{1000}$

f. $4\dfrac{2}{1000}$

g. $17\dfrac{70}{100}$

Lesson 26

Part 5

Answer each question.

	Doe Lake	Arrow Lake	Total for both lakes
September	18	56	74
October	34	64	98
Total for both months	52	120	172

Fish Caught at Two Lakes

Part 5	
a.	b.

a. In which month were more fish caught in Arrow Lake?

b. How many fish were caught in Doe Lake during October?

c. In which lake were fewer fish caught in September?

d. How many fish were caught in Arrow Lake during both months?

e. In which month was the total greater for the number of fish caught in both lakes?

Part 6

Write each fraction as a division problem. Figure out the mixed number it equals.

a. $\dfrac{23}{2}$ b. $\dfrac{43}{5}$ c. $\dfrac{73}{9}$ d. $\dfrac{31}{8}$ e. $\dfrac{59}{8}$

Part 7

Copy and work each problem. Rewrite each whole number as a fraction.

a. $\dfrac{15}{8}$
   $- 1$

b. 7
   $- \dfrac{4}{5}$

c. 10
   $- \dfrac{7}{2}$

d. $\dfrac{3}{4}$
   $+ 6$

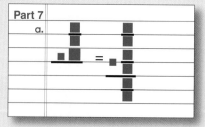

Part 8

Write the answer to each division fact.

a. $9\overline{)72}$ b. $6\overline{)36}$ c. $9\overline{)54}$ d. $7\overline{)28}$ e. $9\overline{)63}$

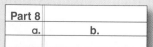

f. $9\overline{)45}$ g. $7\overline{)56}$ h. $8\overline{)48}$ i. $7\overline{)49}$ j. $8\overline{)32}$

Lesson 26

Part 9 Write the **x** and **y** equation for each point.

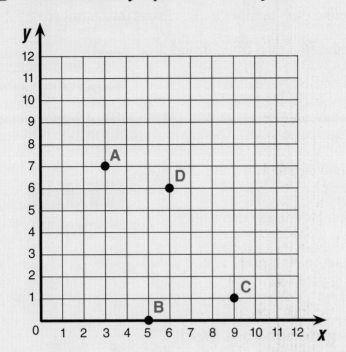

Part 9	
A.	$x = $ ■, $y = $ ■

Part 10 Copy and work each problem.

a. 40
   × 3 1

b. 16
   × 3 3

c. 25
   × 3 5

d. 103
   × 29

Part 10	
a.	▬
	× ▬

Lesson 27

Part 1 Copy each number. Round each number to the closest tens number.

a. 114   b. 296   c. 125   d. 74   e. 85   f. 34

Part 2 Work each problem. Use initials for the names.

a. The ratio of boys to girls at a show is 5 to 4. If there are 32 girls, how many boys are there?

b. Every 2 days, Sarah jogs 8 miles. How many miles does she jog in 10 days?

c. The ratio of windows to rooms is 7 to 1. If there are 63 windows, how many rooms are there?

Independent Work

Part 3 Copy and work each problem.

a.
$$\frac{15}{10}$$
$$-\frac{4}{5}$$

b.
$$\frac{6}{4}$$
$$-\frac{10}{7}$$

c.
$$\frac{5}{6}$$
$$+\frac{8}{9}$$

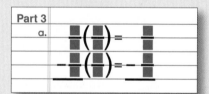

Part 4 Answer each question.

	Red cars	Blue cars	Total for both cars
Hill Street	48	63	111
Vale Street	38	65	103
Total for both streets	86	128	214

Cars on Two Streets

a. What is the total number of red and blue cars on Vale Street?

b. How many blue cars are on Hill Street?

c. Are there more blue cars on Hill Street or Vale Street?

d. What is the total number of red cars for both streets?

e. Are there more red cars or blue cars on Hill Street?

Lesson 27

Part 5 Write the decimal value for each mixed number.

a. $1\frac{4}{10}$ b. $22\frac{7}{100}$ c. $3\frac{4}{1000}$ d. $29\frac{10}{1000}$

Part 6 Work each problem. Remember the unit name.

a. King Mountain is 8700 feet above sea level. Anchor Mountain is 3620 feet higher than King Mountain. What is the height of Anchor Mountain?

b. Rover ran the course in 17 seconds less than Dover. Dover took 231 seconds to run the course. How many seconds did it take Rover to run the course?

c. The land turtle was 97 years younger than the sea turtle. The land turtle was 9 years old. How old was the sea turtle?

Part 6	
a.	

Part 7 Write the answer to each division fact.

a. $9\overline{)63}$ b. $8\overline{)48}$ c. $3\overline{)24}$ d. $5\overline{)45}$ e. $9\overline{)36}$

f. $7\overline{)42}$ g. $4\overline{)24}$ h. $9\overline{)63}$ i. $9\overline{)54}$ j. $6\overline{)30}$

Part 7	
a.	b.

Part 8 Copy and work each problem.

a. $\begin{array}{r} 34 \\ \times 92 \\ \hline \end{array}$
b. $\begin{array}{r} 76 \\ \times 35 \\ \hline \end{array}$
c. $\begin{array}{r} 47 \\ \times 90 \\ \hline \end{array}$
d. $\begin{array}{r} 94 \\ \times 48 \\ \hline \end{array}$

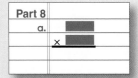

Lesson 28

Part 1
Change the whole number into a simple fraction and multiply.

a. $8 \times \dfrac{7}{3} = \blacksquare$ b. $\dfrac{5}{6} \times 9 = \blacksquare$ c. $2 \times \dfrac{9}{12} = \blacksquare$ d. $\dfrac{5}{4} \times 11 = \blacksquare$

Part 2
Copy each number. Then write the closest tens number.

a. 404   b. 76   c. 233   d. 165

e. 614   f. 305   g. 24   h. 145

Independent Work

Part 3
Work each problem. Use initials for the names.

a. There were 7 bottles in every case. There were 84 bottles. How many cases were there?

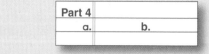

b. Each circle had 9 parts. There were 46 circles. How many parts were there?

c. Every bag of feathers weighed 8 ounces. The weight of all the bags was 112 ounces. How many bags were there?

Part 4
Answer each question.

	Lunches	Dinners	Total for both meals
Tuesday	461	143	604
Wednesday	372	317	689
Total for both days	833	460	1293

Meals Served on Two Days

a. How many dinners were served on Tuesday?

b. What was the total number of lunches served on both days?

c. On Wednesday, were more lunches or dinners served?

d. What was the total number of meals served on both days?

e. Were fewer dinners served on Tuesday or on Wednesday?

Connecting Math Concepts

Lesson 28

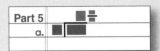

Part 5 Write each fraction as a division problem. Figure out the whole number or mixed number it equals.

a. $\dfrac{435}{4}$ b. $\dfrac{712}{6}$ c. $\dfrac{501}{8}$

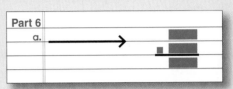

Part 6 Make a number family for each item. Figure out the missing number.

a. The big number is K. The small numbers are F and 15.

$$K = 220$$

b. The big number is 187. The small numbers are B and E.

$$E = 19$$

Part 7 Copy and work each problem.

a. $\begin{array}{r} 56 \\ \times\ 35 \\ \hline \end{array}$
b. $\begin{array}{r} 24 \\ \times\ 72 \\ \hline \end{array}$
c. $\begin{array}{r} 83 \\ \times\ 46 \\ \hline \end{array}$
d. $\begin{array}{r} 55 \\ \times\ 38 \\ \hline \end{array}$

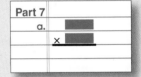

Part 8 Write the answer to each division fact.

a. $2\overline{)16}$ b. $4\overline{)28}$ c. $6\overline{)36}$ d. $8\overline{)64}$ e. $5\overline{)35}$

f. $4\overline{)32}$ g. $9\overline{)72}$ h. $8\overline{)56}$ i. $7\overline{)21}$ j. $8\overline{)32}$

Part 9 Copy and work each problem. Rewrite each whole number as a fraction.

a. $\begin{array}{r} 9 \\ +\ \frac{2}{5} \\ \hline \end{array}$
b. $\begin{array}{r} 5 \\ -\ \frac{3}{10} \\ \hline \end{array}$
c. $\begin{array}{r} 4 \\ +\ \frac{9}{6} \\ \hline \end{array}$
d. $\begin{array}{r} \frac{12}{5} \\ -\ 1 \\ \hline \end{array}$

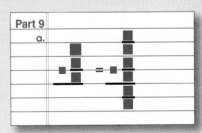

Lesson 29

Part 1

Work each problem. Use initials for the names.

a. There were 24 cans in each bag. There were 9 bags. How many cans were there?

b. Every machine weighed 9 tons. All the machines weighed 108 tons. How many machines were there?

c. There were 6 people in each boat. There were 24 boats. How many people were there?

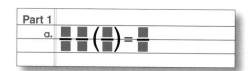

Part 2

Write each fraction multiplication problem and work it.

a. $\frac{3}{2}$ of 20   b. $\frac{7}{8}$ of 40   c. $\frac{1}{2}$ of 28   d. $\frac{2}{3}$ of 9

Independent Work

Part 3

Answer each question.

	White pelicans	Brown pelicans	Total for both colors
Peat Island	464	587	1051
Haystack Rock	258	804	1062
Total for both refuges	722	1391	2113

Pelicans Seen at Two Wildlife Refuges

Part 3	
a.	b.

a. How many white pelicans were seen at Haystack Rock?

b. What's the total number of brown pelicans seen at both refuges?

c. At which refuge were fewer white pelicans seen?

d. What's the total number of pelicans seen at both refuges?

e. Were more pelicans seen at Peat Island or Haystack Rock?

Lesson 29

Part 4 Write the decimal value for each mixed number.

a. $40\frac{4}{100}$ b. $16\frac{3}{10}$ c. $58\frac{26}{1000}$ d. $29\frac{80}{1000}$

Part 5 Copy and work each problem.

a.
$$\frac{5}{8}$$
$$+\frac{4}{16}$$

b.
$$\frac{3}{6}$$
$$+\frac{2}{3}$$

c.
$$\frac{25}{8}$$
$$-\frac{12}{4}$$

d.
$$\frac{20}{4}$$
$$-\frac{2}{3}$$

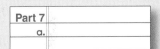

Part 6 Copy and work each problem.

a.
$$\begin{array}{r} 40 \\ \times\ 76 \\ \hline \end{array}$$

b.
$$\begin{array}{r} 37 \\ \times\ 58 \\ \hline \end{array}$$

c.
$$\begin{array}{r} 93 \\ \times\ 46 \\ \hline \end{array}$$

d.
$$\begin{array}{r} 89 \\ \times\ 40 \\ \hline \end{array}$$

Part 7 Work each problem. Remember the unit name.

a. The boat weighed 3614 pounds less than the truck. The truck weighed 9609 pounds. How much did the boat weigh?

b. In March the cow weighed 468 pounds. In June the cow weighed 667 pounds. How much weight did the cow gain?

c. In the morning, the store had 1880 packages of napkins. In the evening, the store had 748 packages of napkins. How many packages did the store sell that day?

Lesson 30

Part 1
Write the decimal value for each fraction.

a. $\dfrac{375}{100}$ b. $\dfrac{375}{10}$ c. $\dfrac{375}{1000}$

d. $\dfrac{1201}{100}$ e. $\dfrac{89}{10}$ f. $\dfrac{52}{100}$

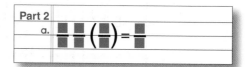

Part 1			
a.		b.	c.
d.		e.	f.

Part 2
Work each problem. Use initials for the names.

a. Every 4 cans weighed 5 pounds. All the cans weighed 200 pounds. How many cans were there?

b. The ratio of cats to mice was 2 to 7. There were 64 cats. How many mice were there?

c. There were 64 teams. Each team had 9 players. How many players were there?

Part 3
Write each fraction multiplication problem and work it.

a. $\dfrac{5}{3}$ of 9   b. $\dfrac{2}{5}$ of 30   c. $\dfrac{4}{3}$ of 90   d. $\dfrac{1}{7}$ of 56

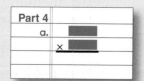

Independent Work

Part 4
Copy and work each problem.

a. $\begin{array}{r} 53 \\ \times\ 38 \\ \hline \end{array}$ b. $\begin{array}{r} 27 \\ \times\ 45 \\ \hline \end{array}$ c. $\begin{array}{r} 36 \\ \times\ 80 \\ \hline \end{array}$ d. $\begin{array}{r} 47 \\ \times\ 53 \\ \hline \end{array}$

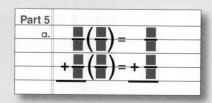

Part 5
Copy and work each problem.

a. $\begin{array}{r} \frac{4}{7} \\ +\ \frac{1}{3} \\ \hline \end{array}$ b. $\begin{array}{r} \frac{11}{8} \\ -\ \frac{3}{4} \\ \hline \end{array}$ c. $\begin{array}{r} \frac{12}{5} \\ +\ \frac{1}{2} \\ \hline \end{array}$ d. $\begin{array}{r} \frac{13}{10} \\ -\ \frac{6}{5} \\ \hline \end{array}$

Lesson 30

Part 6 Write the answer to each division fact.

a. $10\overline{)60}$ b. $5\overline{)20}$ c. $6\overline{)24}$ d. $7\overline{)56}$

e. $5\overline{)40}$ f. $6\overline{)42}$ g. $5\overline{)35}$ h. $6\overline{)48}$

Part 6		
a.	b.	

Part 7 Write each fraction as a division problem. Figure out the whole number or mixed number it equals.

a. $\dfrac{276}{9}$ b. $\dfrac{350}{8}$ c. $\dfrac{129}{6}$

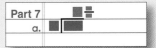

Part 7		
a.		

Lesson 31

Part 1 Write the decimal value for each fraction.

a. $\dfrac{62}{100}$ b. $\dfrac{504}{10}$ c. $\dfrac{2158}{1000}$ d. $\dfrac{1823}{100}$ e. $\dfrac{98}{10}$ f. $\dfrac{709}{100}$

Part 1	
a.	

Part 2 Write each fraction multiplication problem and work it.

a. $\dfrac{1}{3}$ of 603   b. $\dfrac{5}{4}$ of 30   c. $\dfrac{2}{5}$ of 25   d. $\dfrac{10}{9}$ of 81

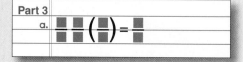

Independent Work

Part 3 Work each problem. Use initials for the first fraction.

a. Each train carried 21 cars of rock. How many cars of rock could 12 trains carry?

b. The ratio of trees to bushes was 7 to 9. There were 180 bushes. How many trees were there?

c. They painted 4 houses every 18 days. They painted a total of 80 houses. How many days did they paint houses?

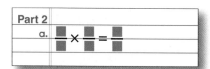

Part 4 Copy and work each problem. Rewrite each whole number as a fraction.

a.
$$\dfrac{4}{5}$$
$$+\dfrac{7}{10}$$

b.
$$\dfrac{12}{6}$$
$$-\dfrac{1}{4}$$

c.
$$\dfrac{4}{7}$$
$$+\;2$$

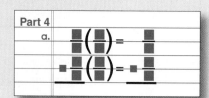

Part 5 Copy and work each problem.

a.
$$\begin{array}{r} 5\,2\,1 \\ \times\ \ 1\,6 \\ \hline \end{array}$$

b.
$$\begin{array}{r} 3\,0\,8 \\ \times\ \ 3\,7 \\ \hline \end{array}$$

c.
$$\begin{array}{r} 4\,8\,3 \\ \times\ \ 5\,0 \\ \hline \end{array}$$

d.
$$\begin{array}{r} 7\,8\,0 \\ \times\ \ \ \ 9 \\ \hline \end{array}$$

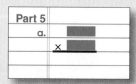

Part 6 Write each fraction as a division problem. Figure out the whole number or mixed number it equals.

a. $\dfrac{547}{5}$ b. $\dfrac{354}{3}$ c. $\dfrac{673}{7}$

Part 6	
a.	

Lesson

Part 1 Write each number.

a. Thirty two thousand one hundred forty.

b. One hundred twelve thousand five hundred four.

c. Seven hundred five thousand thirty two.

d. Fifty six thousand three hundred.

e. Two hundred thousand nine hundred ninety one.

Part 2 Work each problem.

a. What's $\frac{3}{5}$ of 25?

b. What's $\frac{4}{7}$ of 10?

c. What's $\frac{2}{3}$ of 20?

d. What's $\frac{1}{4}$ of 36?

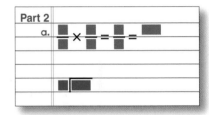

Independent Work

Part 3 Write each fraction as a division problem. Figure out the whole number or mixed number it equals.

a. $\frac{381}{2}$

b. $\frac{577}{5}$

c. $\frac{381}{7}$

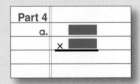

Part 4 Copy and work each problem.

a. $\begin{array}{r} 39 \\ \times 17 \\ \hline \end{array}$

b. $\begin{array}{r} 36 \\ \times 84 \\ \hline \end{array}$

c. $\begin{array}{r} 72 \\ \times 44 \\ \hline \end{array}$

d. $\begin{array}{r} 508 \\ \times 41 \\ \hline \end{array}$

Part 5 Write the decimal value that equals each fraction.

a. $\frac{58}{10}$

b. $\frac{42}{100}$

c. $\frac{2560}{1000}$

d. $\frac{7}{10}$

e. $\frac{101}{100}$

Lesson 32

Part 6 Work each problem. Use initials for the first fraction.

a. The ratio of bottles to cans was 11 to 9. There were 189 cans. How many bottles were there?

b. There were 5 campers in each tent. They put up 60 tents. How many campers were there in all?

c. Every 3 packages cost $5. The cost of all the packages was $40. How many packages were there?

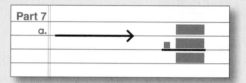

Part 7 Make a number family. Figure out the missing number.

a. P is 11 less than K. K is 34. What number is P?

b. T is 15 more than Y. T is 81. What number is Y?

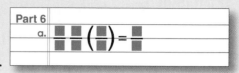

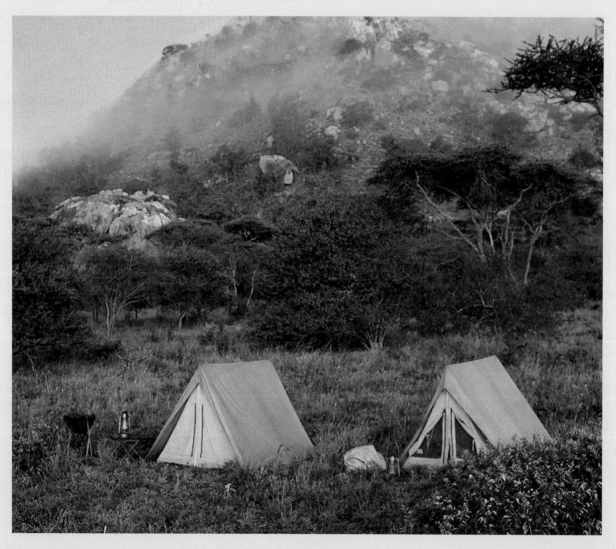

Connecting Math Concepts

Lesson 33

Part 1 Write each number.

   a. One hundred thousand fifty six.

   b. Fifty six thousand two hundred.

   c. One hundred eight thousand seven.

   d. Twenty thousand six hundred ninety.

   e. Three hundred ninety thousand.

Part 2 Write the coordinates for each point.

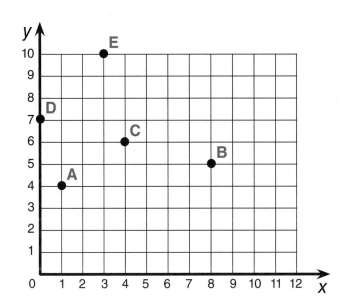

Part 3 Work each problem.

  a. What's $\dfrac{5}{2}$ of 20?

  b. What's $\dfrac{3}{5}$ of 11?

  c. What's $\dfrac{1}{4}$ of 56?

  d. What's $\dfrac{2}{3}$ of 40?

Lesson 33

Independent Work

Part 4 Copy and work each problem.

a. 48
   × 84

b. 109
   × 98

c. 562
   × 34

d. 693
   × 23

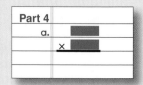

Part 5 Work each problem. Use initials for the first fraction.

a. The cost of every 3 brushes is 10 dollars. The cost of all the brushes is 80 dollars. How many brushes are there?

b. There were 12 packages in each carton. There were 8 cartons. How many packages were there?

c. The ratio of workers to machines was 5 to 3. There were 36 machines. How many workers were there?

Part 6 Write each fraction as a division problem. Figure out the whole number or mixed number it equals.

a. $\dfrac{249}{6}$

b. $\dfrac{473}{9}$

c. $\dfrac{803}{6}$

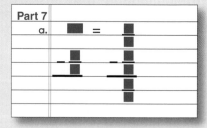

Part 7 Copy and work each problem. Rewrite each whole number as a fraction.

a. 6
   $-\dfrac{3}{4}$

b. $\dfrac{2}{3}$
   $+\dfrac{1}{9}$

c. $\dfrac{3}{8}$
   $+\dfrac{5}{6}$

Lesson 34

Part 1
Write the coordinates for each point.

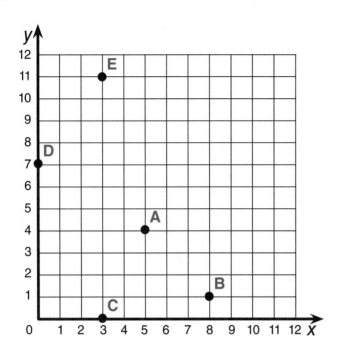

Part 1	
A.	(■, ■)

Part 2
Work each problem.

a. What's $\frac{8}{7}$ of 49?

b. What's $\frac{3}{2}$ of 21?

c. What's $\frac{1}{9}$ of 90?

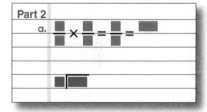

Independent Work

Part 3
Work each problem. Use initials for the first fraction.

a. The ratio of children to adults was 3 to 4. There were 36 children. How many adults were there?

b. Each ribbon is 10 inches long. The total length of all the ribbons is 240 inches. How many ribbons are there?

c. There were 6 fountains in every mall. There were 19 malls. How many fountains were there?

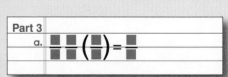

Lesson 34

Part 4   Copy and work each problem.

a.
```
  565
×   8
```

b.
```
  197
× 20
```

c.
```
  36
×63
```

d.
```
  590
× 75
```

Part 5   Write each fraction as a division problem. Figure out the whole number or mixed number it equals.

a. $\dfrac{356}{5}$

b. $\dfrac{748}{7}$

c. $\dfrac{129}{8}$

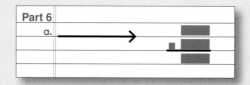

Part 6   Make a number family. Figure out the missing number.

a. R is 52 less than C. C is 106. What number is R?

b. J is 44 more than T. J is 205. What number is T?

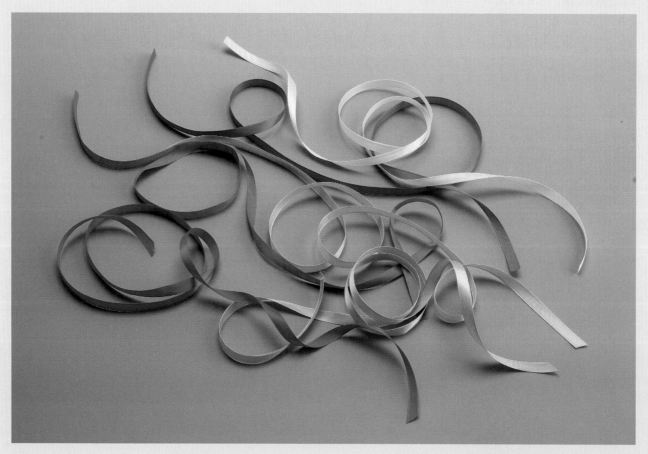

Lesson

Part 1 Work each problem. Use initials for the first fraction.

a. There were 11 lakes for every 3 mountains. There were 57 mountains. How many lakes were there?

b. The ratio of cups to saucers was 7 to 4. There were 56 cups. How many saucers were there?

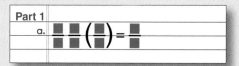

Part 2 Copy and work each problem.

a. 3 5 4
  × 2 2

b. 7 6
  × 3 1

c. 4 1 8
  ×   5 0

d. 7 2 0
  ×   3 8

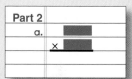

Part 3 Write the coordinates for each point.

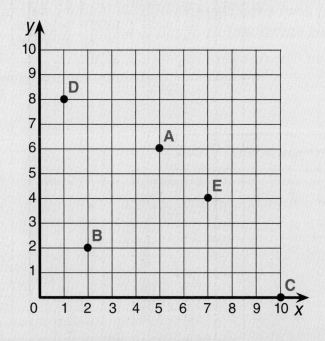

Part 4 Copy and work each problem.

a. $\dfrac{2}{7}$
  $+\dfrac{3}{2}$

b. $\dfrac{5}{4}$
  $+\dfrac{1}{2}$

c. $\dfrac{8}{10}$
  $-\dfrac{4}{5}$

Lesson 36

Part 1
Write each number. Remember the commas.

a. 17 million 230 thousand ten.

b. 1 million 289 thousand 6 hundred 7.

c. 4 million 20 thousand 4 hundred.

d. 560 million 8 thousand fifty.

e. 2 million 30 thousand 8 hundred 4.

Part 2
Multiply each fraction by its reciprocal.

a. $\dfrac{3}{8}$ b. $\dfrac{7}{50}$ c. $\dfrac{9}{2}$

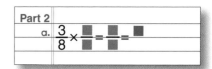

Part 3
Work each problem.

a. There were 4 old cards for every 3 new cards. <u>There were 42 cards. How many old cards were there?</u>

b. For every 10 burgers, 7 were frozen. The rest were thawed. If there were 21 thawed burgers, how many burgers were frozen?

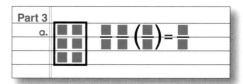

Independent Work

Part 4
Write each problem in a column and work it.

a. $10 + \dfrac{7}{3}$ b. $\dfrac{2}{5} + \dfrac{8}{2}$

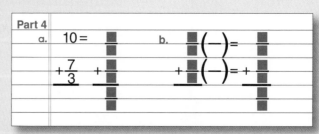

Part 5
Write the reciprocal of each fraction.

a. $\dfrac{1}{40}$ b. $\dfrac{32}{15}$

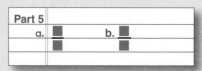

Lesson 36

Part 6 Copy and work each problem.

a.
$$\begin{array}{r} 83 \\ \times\,40 \\ \hline \end{array}$$

b.
$$\begin{array}{r} 126 \\ \times\ \ 16 \\ \hline \end{array}$$

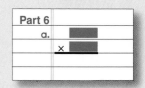

Part 7 Write the coordinates for each point.

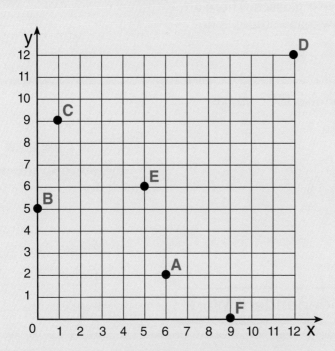

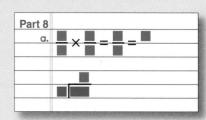

Part 8 Work each problem. Write the answer as a whole number or mixed number.

a. What's $\dfrac{2}{3}$ of 78?

b. What's $\dfrac{3}{7}$ of 55?

Lesson 37

Part 1 Work each problem.

a. The ratio of empty bottles to full bottles was 7 to 1. If there were 9 full bottles how many bottles were there in all?

b. There are green balloons and red balloons at a party. The ratio of green balloons to red balloons is 3 to 5. If there are 240 balloons, how many balloons are red?

c. For every 7 students, 2 students wear glasses. There are 64 students who wear glasses. How many students do not wear glasses?

Part 2 Write each number. Remember the commas.

a. 20 million 5 thousand 2 hundred 10.

b. 6 million 12 thousand twenty.

c. 108 million 108 thousand 1 hundred 8.

d. 15 million 40 thousand 7 hundred 3.

e. 2 million 600 thousand 42.

Independent Work

Part 3 Copy and work each problem.

a.   46
   × 3 2

b.   3 0 2
   ×   5 0

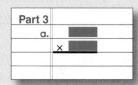

Part 4 Copy and work each problem. Show the answer as a mixed number.

a. 5⟌4 6   b. 3⟌1 9   c. 9⟌8 8

Lesson 37

Part 5 Write the coordinates for each point.

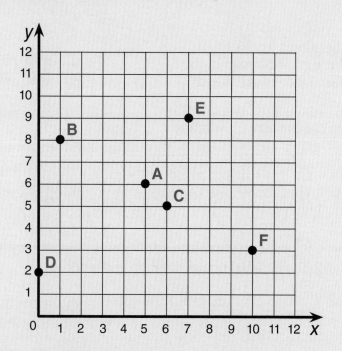

Part 5	
A.	(■, ■)

Part 6 Work each problem.

a. T is 208. T is 16 less than J. What number is J?

b. P is 136 more than K. P is 206. What number is K?

Part 6	
a.	

Part 7 Write each fraction as a division problem. Figure out the whole number or mixed number it equals.

a. $\dfrac{408}{3}$

b. $\dfrac{76}{9}$

Part 7	
a.	

Lesson

Part 1   Work each problem.

a. In a study room, the ratio of hardback books to paperback books is 5 to 4. There are 72 books in the study room. How many hardback books are there?

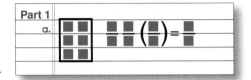

b. After the storm, there were 6 broken windows for every 10 windows. 20 windows were not broken. How many windows were there in all?

c. Joe collects quarters and dimes. He has 4 quarters for every 3 dimes. He has 96 dimes. How many coins does he have in his collection?

Part 2   Write each problem with the decimal points lined up and work it.

a. $4.1 + 3 + 12.05$ b. $23.2 + .16 + 8$ c. $6 + .7 + .001$

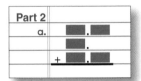

Independent Work

Part 3   Write each problem in a column and work it.

a. $7 - \dfrac{5}{3}$ b. $\dfrac{7}{4} + \dfrac{3}{6}$

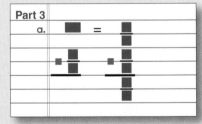

Part 4   Work each problem. Write the answer as a whole number or mixed number.

a. What's $\dfrac{3}{5}$ of 98?   b. What's $\dfrac{2}{9}$ of 81?

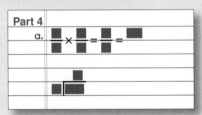

Lesson 38

Part 5 Write each number.

a. 8 million 205 thousand one.

b. 3 million 10 thousand 6 hundred.

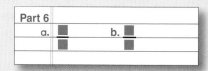

Part 6 Write the reciprocal of each fraction.

a. $\dfrac{13}{8}$ b. $\dfrac{4}{59}$

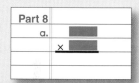

Part 7 Copy and work each problem. Show the answer as a mixed number.

a. $7\overline{)59}$ b. $9\overline{)59}$ c. $8\overline{)49}$

Part 7

Part 8 Copy and work each problem.

a. $\begin{array}{r} 108 \\ \times\ 23 \\ \hline \end{array}$ b. $\begin{array}{r} 235 \\ \times\ 50 \\ \hline \end{array}$

Part 8

Lesson 39

Part 1 Copy and work each problem.

a. $55\overline{)308}$ with 6 above

b. $25\overline{)170}$ with 6 above

c. $78\overline{)498}$ with 7 above

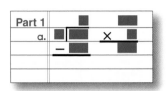

Part 2 Work each problem.

a. On a ship there were 2 crew members for every 10 passengers. There were 600 passengers on the ship. How many crew members were on the ship?

b. On another ship there were 6 women for every 5 men. There were 80 men on the ship. How many adults were on the ship?

c. The ratio of new buildings to old buildings in a neighborhood was 4 to 7. There were 200 new buildings. How many total buildings were in the neighborhood?

d. The ratio of cows to all animals on a farm was 5 to 9. There were 72 animals on the farm. How many cows were on the farm?

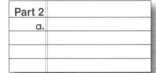

Independent Work

Part 3 Write the coordinates for each point.

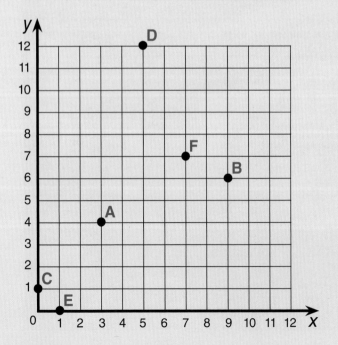

Part 3
A. (■, ■)

Lesson 39

Part 4 Write each number.

a. 5 million 760 thousand thirteen.

b. 2 million 4 thousand 24.

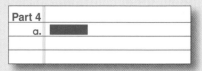

Part 5 Work each problem. Write the answer as a whole number or mixed number.

a. What's $\frac{4}{3}$ of 65?

b. What's $\frac{4}{3}$ of 30?

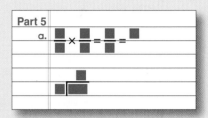

Part 6 Work each problem.

a. R is 112 more than V. R is 127. What number is V?

b. B is 168. K is 98 more than B. What number is K?

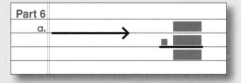

Part 7 Write each fraction as a division problem. Figure out the whole number or mixed number it equals.

a. $\frac{78}{3}$

b. $\frac{794}{6}$

Lesson 40

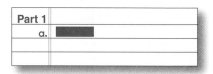

Part 1 Write the place value of each arrowed digit.

a. 7,894,312
↑

b. 7,894,312
↑

c. 7,894,312
↑

Part 2 Work each problem.

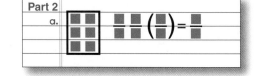

a. On a lot, the ratio of new cars to all cars was 3 to 10. There were 90 cars on the lot. How many were used cars?

b. There were 9 bees for every 2 flowers. There were 18 flowers. How many bees were there?

c. The company buys 3 red boxes for every 7 white boxes. The company bought a total of 670 boxes last week. How many were red boxes?

d. The recipe uses 3 pints of milk for every 12 eggs. How many eggs are needed for 9 pints of milk?

Independent Work

Part 3 Copy and work each problem. If the answer shown is too large, work the problem again with an answer that is 1 less.

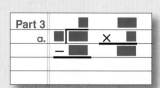

a. 42⟌323 (8)

b. 65⟌296 (4)

c. 84⟌250 (3)

Part 4 Write each problem with the decimal points lined up and work it.

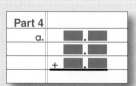

a. 5.016 + 3.1 + 18.2

b. 15.60 − 3.09

c. 8.7 + 10 + .045

Lesson 40

Part 5 | Write the reciprocal of each fraction.

a. $\dfrac{2}{28}$

b. $\dfrac{46}{1}$

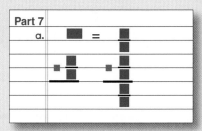

Part 6 | Copy and work each problem.

a.
$$\begin{array}{r} 4\,7 \\ \times\,8\,2 \\ \hline \end{array}$$

b.
$$\begin{array}{r} 5\,2\,5 \\ \times\quad\ 5 \\ \hline \end{array}$$

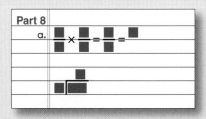

Part 7 | Write each problem in a column and work it.

a. $8 - \dfrac{8}{3}$

b. $\dfrac{15}{9} - \dfrac{2}{3}$

Part 8 | Work each problem. Write the answer as a whole number or mixed number.

a. What's $\dfrac{9}{5}$ of 60?

b. What's $\dfrac{2}{7}$ of 48?

Part 9 | Copy and work each problem. Show the answer as a mixed number.

a. $7\overline{)6\,8}$

b. $8\overline{)6\,8}$

c. $6\overline{)5\,6}$

Lesson 41

Part 1
Write the letter of each parallelogram.

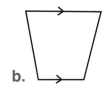

a.

b.

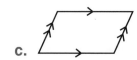

c.

d.

e.

Part 1	

Part 2
Multiply and subtract. If the remainder is too big, work the problem again with an answer that is 1 more.

a. $17\overline{)120}$ with 6 above

b. $41\overline{)214}$ with 4 above

c. $38\overline{)335}$ with 8 above

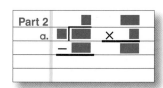

Part 3
Write the place value of each arrowed digit.

a. 6,213,897 ↑ (under the 3)

e. 4,837,512 ↑ (under the 7)

b. 6,213,897 ↑ (under the 6)

f. 4,837,512 ↑ (under the 5)

c. 6,213,897 ↑ (under the 1)

g. 4,837,512 ↑ (under the 8)

d. 6,213,897 ↑ (under the 3)

h. 4,837,512 ↑ (under the 4)

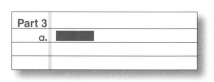

Part 4
Copy and work each problem.

a.
```
   4 0 6
×    .3 3
```

b.
```
   1.3 5
×     .4
```

c.
```
   2.5
× 3.0
```

d.
```
   5.0 7
×     .9
```

Lesson 41

Independent Work

Part 5 Write each problem in a column and work it.

a. $\dfrac{16}{3} - \dfrac{11}{9}$ b. $\dfrac{18}{5} - \dfrac{5}{3}$

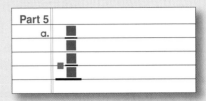

Part 6 Copy and work each problem.

a. $\begin{array}{r} 6\,1\,6 \\ \times\ \ 3\,1 \end{array}$ b. $\begin{array}{r} 4\,6 \\ \times\ 8\,9 \end{array}$ c. $\begin{array}{r} 2\,7\,8 \\ \times\ \ 8\,4 \end{array}$

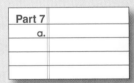

Part 7 Work each problem. Use initials for the first fraction.

a. There were 6 rainy days for every 9 days. There were 54 days. How many sunny days were there?

b. There were 64 children on the track team. There were 5 girls for every 3 boys on the team. How many girls were on the team?

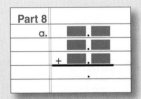

Part 8 Write each problem with the decimal points lined up and work it.

a. 12.06 + 281 + .765   b. 56.017 − 8.1

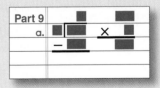

Part 9 Copy and work each problem. If the answer shown is too large, work the problem again with an answer that is 1 less.

a. $23\overline{)191}$ with 9 above   b. $51\overline{)434}$ with 8 above   c. $74\overline{)661}$ with 9 above

Lesson 42

Part 1 Copy and work each problem.

a. $41\overline{)246}$ (quotient 5)

b. $65\overline{)260}$ (quotient 3)

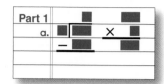

Part 2 Copy each item and figure out what the letter equals.

a. $\dfrac{7}{5}$ M = 14

     M = ■ = ■

b. $\dfrac{2}{9}$ L = 40

     L = ■ = ■

c. $\dfrac{1}{5}$ J = 10

     J = ■ = ■

d. $\dfrac{2}{3}$ B = 24

     B = ■ = ■

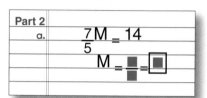

Part 3 Copy and work each problem.

a.
```
  903
 -113
```

b.
```
  108
 - 29
```

c.
```
  501
 -145
```

d.
```
  400
 -150
```

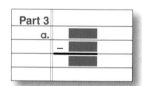

Independent Work

Part 4 Copy and work each problem. Show the answer as a mixed number.

a. $6\overline{)56}$ b. $6\overline{)51}$ c. $9\overline{)60}$ d. $9\overline{)33}$

Part 5 Write each fraction as a division problem. Figure out the whole number or mixed number it equals.

a. $\dfrac{148}{6}$ b. $\dfrac{472}{5}$

Connecting Math Concepts

Part 6 Work each problem. Write the answer as a whole number or a mixed number.

a. What's $\frac{3}{5}$ of 90?

b. What's $\frac{7}{8}$ of 80?

c. What's $\frac{2}{3}$ of 66?

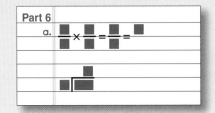

Part 7 Copy and work each problem. If the answer shown is too large, work the problem again with an answer that is 1 less.

a. $54\overline{\smash)343}$ ⁶

b. $78\overline{\smash)463}$ ⁶

Part 8 Write the place value for the arrowed digit.

a. 2,340,567
↑

b. 4,079,300
↑

c. 3,345,671
↑

Lesson 43

Part 1

Find the area of each parallelogram. Remember the unit name in the answer.

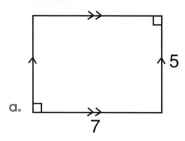

a.

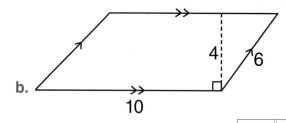

b.

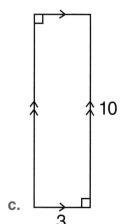

c.

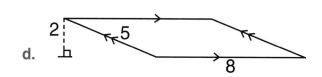

d.

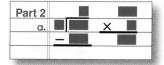

Part 1	
a.	■ × ■ = ▨▨

Part 2

Copy and work each problem.

a. $58\overline{)464}$ ⁷

b. $72\overline{)373}$ ⁴

c. $45\overline{)409}$ ⁸

d. $13\overline{)78}$ ⁵

Part 3

Round each number to the nearest thousand.

a. 2468   b. 8602   c. 1500   d. 6356

Part 3	
a.	▨

Part 4

Write each problem in a column and work it.

a. .203 × .03   b. .018 × .9

c. 5.10 × .21   d. .14 × .006

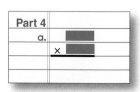

Lesson 43

Part 5 Copy each item and figure out what the letter equals.

a. $\dfrac{3}{4}$ R = 15

  R = ▮ = ■

b. $\dfrac{1}{3}$ T = 7

  T = ▮ = ■

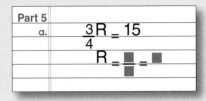

c. $\dfrac{5}{2}$ M = 10

  M = ▮ = ■

d. $\dfrac{2}{7}$ F = 20

  F = ▮ = ■

Part 6 Copy and work each problem. If the answer shown is too large, work the problem again with an answer that is 1 less.

a. $63\overline{)491}$ with 8 above

b. $34\overline{)271}$ with 7 above

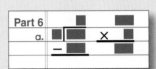

Part 7 Work each problem. Use initials for the first fraction.

a. There were crows and sparrows in the field. There were 3 crows for every 5 sparrows. There were 56 birds in the field. How many were sparrows?

b. The ratio of old shoes to all the shoes in the closet was 7 to 11. There were 21 old shoes. How many new shoes were there?

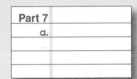

Part 8 Work each problem.

a. T is 401. B is 138 less than T. What number is B?

b. P is 113 less than M. M is 148. What number is P?

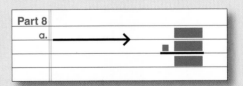

Lesson 43

Part 9 Write the place value for the arrowed digit.

a. 9,740,673
b. 2,806,274
c. 4,855,316

(arrows pointing up under the digits)

Part 9
a.	

Part 10 Write each problem in a column and work it.

a. $\frac{12}{7} - 1$

b. $\frac{3}{8} + \frac{5}{2}$

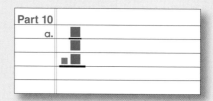

Part 10
a.	

Part 11 Copy and work each problem.

a.
```
   89
×  17
```

b.
```
  458
×  67
```

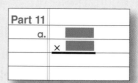

Part 11
a.	
	×

Part 12 Copy and work each problem.

a.
```
  409
−  35
```

b.
```
  607
−  88
```

c.
```
  450
−278
```

d.
```
  703
−178
```

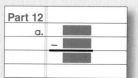

Part 12
a.	
	−

Lesson 44

Part 1 Find the area for each parallelogram. Write sq for square.

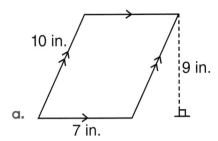

a. 10 in. 9 in. 7 in.

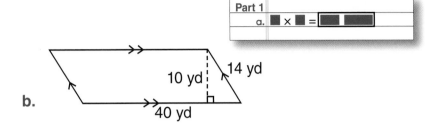

b. 10 yd 14 yd 40 yd

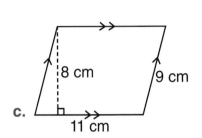

c. 8 cm 9 cm 11 cm

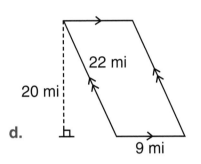

d. 22 mi 20 mi 9 mi

Part 1
a. ■ × ■ = ■ ■

Part 2 Copy each problem. Figure out what each letter equals.

a. $\dfrac{5}{3}(12) = B$ b. $\dfrac{1}{4}(36) = M$ c. $\dfrac{2}{5}(10) = G$

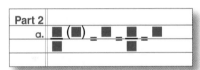

Part 2
a. $\dfrac{■}{■}(■) = \dfrac{■}{■} = \dfrac{■}{■} = ■$

Part 3 Round each number to the closest thousands number.

a. 5467   b. 9601   c. 7520   d. 8356   e. 4700

Part 3
a. ▬

Independent Work

Part 4 Work each problem. Write the answer as a whole number or a mixed number.

a. What's $\dfrac{7}{8}$ of 42?

b. What's $\dfrac{3}{5}$ of 5?

c. What's $\dfrac{7}{9}$ of 63?

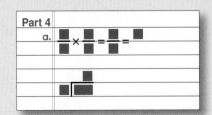

Part 4
a. $\dfrac{■}{■} × \dfrac{■}{■} = \dfrac{■}{■} = ■$

Lesson 44

Part 5

Write the place value for the arrowed digit.

a. 3,615,288   b. 5,410,362   c. 2,565,300

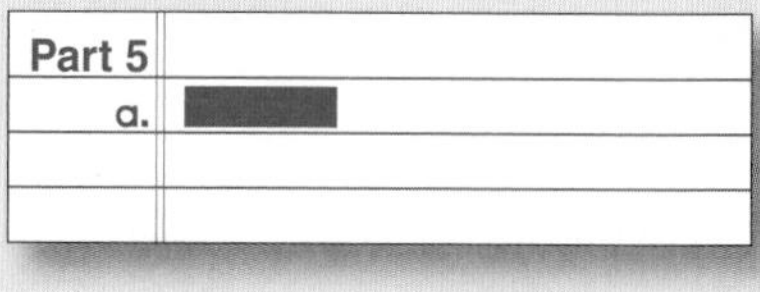

Part 6

Copy and work each problem.

a. $\begin{array}{r} 607 \\ -198 \\ \hline \end{array}$ b. $\begin{array}{r} 403 \\ -\ \ 89 \\ \hline \end{array}$ c. $\begin{array}{r} 906 \\ -245 \\ \hline \end{array}$

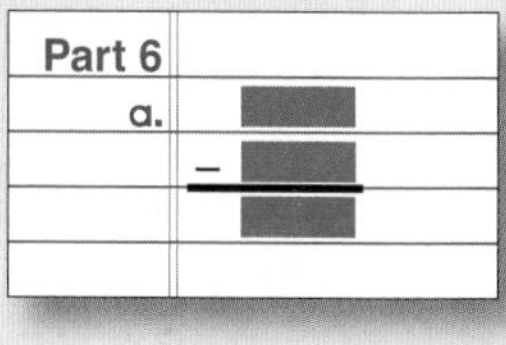

Part 7

Write each fraction as a division problem. Figure out the whole number or mixed number it equals.

a. $\frac{234}{6}$ b. $\frac{210}{5}$

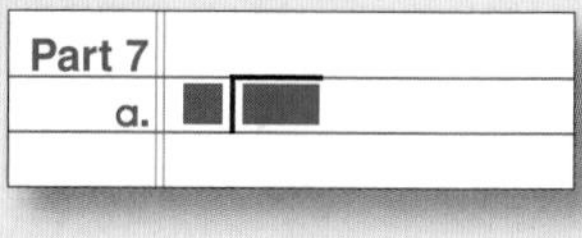

Part 8

Copy and work each problem.

a. $\begin{array}{r} .32 \\ \times\ .08 \\ \hline \end{array}$ b. $\begin{array}{r} 34.2 \\ \times\ \ \ 2.2 \\ \hline \end{array}$ c. $\begin{array}{r} .106 \\ \times\ \ \ \ .5 \\ \hline \end{array}$

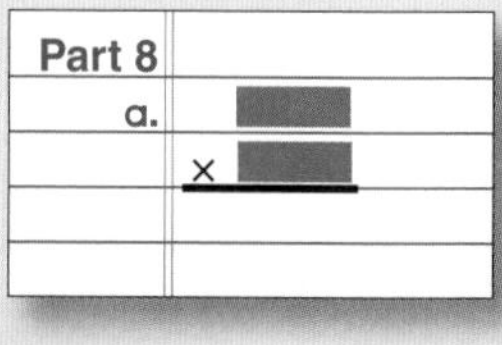

Lesson 45

Part 1 Work each problem.

a. Dolly walked $\frac{5}{8}$ mile to work. Hilary walked $\frac{6}{8}$ mile to work. How much farther did Hilary walk than Dolly walked?

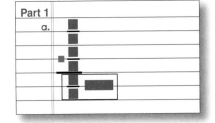

b. The turtle ate $\frac{17}{3}$ pounds of food. The pike ate $\frac{5}{3}$ pounds more than the turtle ate. How many pounds of food did the pike eat?

c. The truck weighed $\frac{22}{10}$ tons. The barge weighed $\frac{35}{10}$ tons. How much heavier was the barge than the truck?

Part 2 Find the area of each figure.

a.
5 yd
4 yd

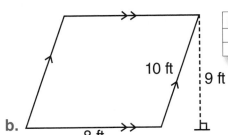

b.
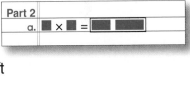
10 ft
9 ft
8 ft

a. ▪ × ▪ = ▪▪▪

c.
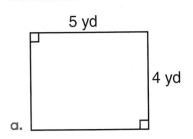
9 in.
3 in. 4 in.

d.
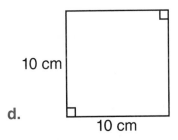
10 cm
10 cm

Part 3 Copy each problem. Figure out what each letter equals.

a. $\frac{5}{9}(18) = F$ b. $\frac{1}{5}T = 9$ c. $\frac{7}{3}(6) = N$

Part 3
a. ▪(▪) = ▪
▪

Lesson 45

Independent Work

Part 4
Work each problem. Write the answer as a whole number or a mixed number.

a. What's $\frac{1}{6}$ of 25? b. What's $\frac{3}{4}$ of 40? c. What's $\frac{3}{5}$ of 99?

Part 5
Copy and work each problem.

a.
```
  .2 6 9
×   .2 1
```

b.
```
  .0 1 8
×   .3 7
```

c.
```
  4.3 6
×   .0 5
```

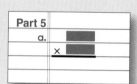

Part 6
Write the place value for the arrowed digit.

a. 8,629,407 ↑   b. 4,525,988 ↑   c. 5,360,480 ↑

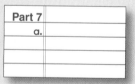

Part 7
Work each problem. Use initials for the first fraction.

a. A farm had turkeys and chickens. There were 2 turkeys for every 3 chickens. There were 50 birds on the farm. How many were chickens?

b. The ratio of cats to dogs in a kennel was 3 to 4. There were 21 cats in the kennel. How many dogs were there?

Part 8
Round each number to the closest thousands number.

a. 2504   b. 1496   c. 6350   d. 4723

Part 9 Copy and work each problem.

a. 902
   −333

b. 506
   −386

c. 605
   −157

Part 9	
a.	▆
−	▆

Part 10 Write the coordinates for each point.

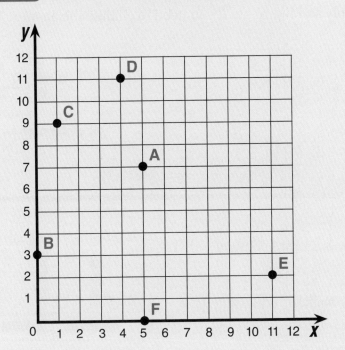

Part 10	
A.	(■, ■)

Lesson 46

Part 1
Round each number to the closest tens number.

a. 199   b. 405   c. 504   d. 713

e. 692   f. 507   g. 294

Part 2
Copy each problem. Multiply and figure out if you need to change the answer. Then work the problem.

a. $23\overline{)161}$ with 6 above

b. $19\overline{)84}$ with 5 above

c. $52\overline{)429}$ with 8 above

d. $47\overline{)191}$ with 3 above

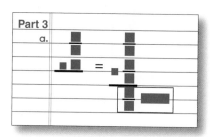

Part 3
Work each problem.

a. The cat was $\frac{9}{4}$ years old. The rat was 2 years younger than the cat. How old was the rat?

b. Will walked $\frac{3}{8}$ mile less than Carly walked. Will walked $\frac{3}{4}$ mile. How far did Carly walk?

c. The paper towels lasted 5 days. The paper towels lasted $\frac{7}{4}$ fewer days than the paper napkins. How many days did the napkins last?

Lesson 46

Part 4 Find the area and perimeter of each figure.

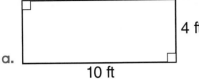

a.

10 ft, 4 ft

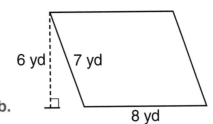

b.

6 yd, 7 yd, 8 yd

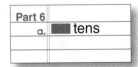

Part 4
a. ▇ × ▇ = ▇▇

Part 5 Write the letter equation for each sentence.

a. $\frac{2}{7}$ of the papers were wet.

b. $\frac{9}{10}$ of the dishes are new.

c. $\frac{3}{8}$ of the people wore glasses.

d. $\frac{4}{5}$ of the bottles were full.

e. $\frac{3}{4}$ of the mice are sleeping.

f. $\frac{2}{3}$ of the coins were expensive.

g. $\frac{6}{7}$ of the papers are graded.

Part 5
a. ▇ ▇ = ▇

Part 6 Write the number of tens for each item.

a. How many tens is 570?

b. How many tens is 500?

c. How many tens is 390?

d. How many tens is 150?

Part 6
a. ▇ tens

Part 7 Copy each problem. Figure out what each letter equals.

a. $\frac{3}{5}R = 15$

b. $\frac{4}{3}m = 20$

c. $\frac{5}{2}(12) = P$

Part 7
a. ▇ ▇ = ▇

Lesson 46

Independent Work

Part 8 — Work each problem.

a. The ratio of red beans to white beans was 5 to 7. There were 120 pounds of beans. How many pounds of white beans were there?

b. On a hill there were 3 small moles for every 2 large moles. There were 48 large moles. How many moles were there?

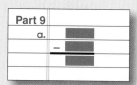

Part 9 — Copy and work each problem.

a.
```
  5 1 3
− 2 4 3
```

b.
```
  7 0 8
− 3 1 1
```

c.
```
  5 9 4
− 4 9 1
```

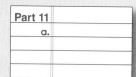

Part 10 — Round each number to the nearest thousand.

a. 6430   b. 81,603   c. 5559   d. 2461   e. 3428

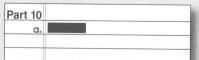

Part 11 — Work each problem. Write the answer as a whole number or a mixed number.

a. What's $\frac{5}{9}$ of 60?

b. What's $\frac{8}{5}$ of 75?

c. What's $\frac{4}{7}$ of 20?

d. What's $\frac{3}{4}$ of 48?

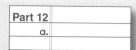

Part 12 — Copy and work each problem.

a.
```
  .3 1 8
×   4.2
```

b.
```
  4.0 3
×  .3 5
```

c.
```
  5.3 2
×  .6 1
```

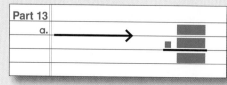

Part 13 — Work each problem.

a. K is 56 less than T. T is 100. What number is K?

b. B is 203 more than K. B is 700. What number is K?

Connecting Math Concepts

Lesson 47

Part 1

Copy each problem. Multiply and figure out if you need to change the answer. Then work the problem.

a. $54\overline{)368}$ with 7

b. $29\overline{)145}$ with 4

c. $63\overline{)327}$ with 5

d. $76\overline{)230}$ with 2

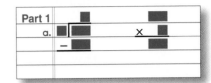

Part 2

Find the area and perimeter of each figure.

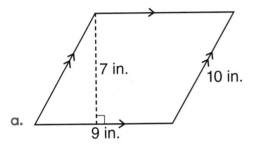

a. (parallelogram) 7 in., 10 in., 9 in.

b. 6 cm, 10 cm

c. (parallelogram) 6 yd, 5 yd, 6 yd

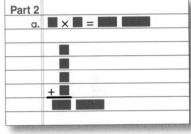

Part 3

Write the letter equation for each sentence.

a. $\frac{4}{10}$ of the books are green.

b. $\frac{1}{8}$ of the pencils were sharpened.

c. $\frac{3}{11}$ of the windows are closed.

d. $\frac{9}{30}$ of the students were finished.

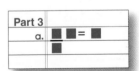

Lesson 47

Part 4
Write the number of tens for each item.

a. How many tens is 80?

b. How many tens is 10?

c. How many tens is 410?

d. How many tens is 280?

e. How many tens is 1200?

Part 4	
a.	■ tens

Part 5
Copy each problem. Replace one of the letters with a number. Figure out the other letter.

a. $\frac{2}{3}J = P$ $\boxed{J = 12}$

b. $\frac{7}{8}B = R$ $\boxed{R = 42}$

c. $\frac{1}{2}M = T$ $\boxed{M = 24}$

d. $\frac{5}{4}V = X$ $\boxed{X = 10}$

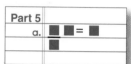

Part 5	
a.	■ ■ = ■
	■

Independent Work

Part 6
Copy and work each problem.

a.
$$712 - 621$$

b.
$$305 - 284$$

c.
$$725 - 527$$

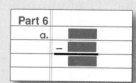

Part 6	
a.	▬
	− ▬
	▬

Part 7
Copy and work each problem.

a.
$$.203 \times .01$$

b.
$$5.42 \times .3$$

c.
$$5.83 \times .04$$

Part 7	
a.	

Part 8
Work each problem. Write the answer as a whole number or a mixed number.

a. What's $\frac{8}{7}$ of 49?

b. What's $\frac{2}{5}$ of 100?

c. What's $\frac{5}{8}$ of 72?

Part 8	
a.	

Lesson 47

Part 9 Work each problem.

a. On the ranch there were 3 old bulls for every 4 young bulls. There were 28 bulls. How many old bulls were there?

b. In a park, the ratio of large ducks to all ducks was 4 to 7. There were 84 ducks. How many large ducks were there?

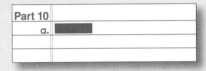

Part 10 Round each number to the nearest thousand.

a. 5050   b. 4683   c. 6530   d. 7390   e. 5349

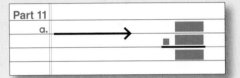

Part 11 Work each problem.

a. J is 16 more than P. P is 14. What number is J?

b. Q is 33 less than M. M is 34. What number is Q?

c. T is 99 less than R. T is 299. What number is R?

Lesson 48

Part 1 — Work each problem.

a. There were some cookies in a jar. Children ate 24 of the cookies. There were still 36 cookies in the jar. How many cookies started out in the jar?

b. There were 46 cookies in a jar. The children ate some cookies. The jar ended up with 31 cookies. How many cookies did the children eat?

c. Sam started out with $\frac{6}{5}$ pounds of grapes. He picked some more grapes. He ended up with $\frac{15}{2}$ pounds of grapes. How many pounds of grapes did he pick?

d. In the morning, the boat sailed $\frac{7}{5}$ miles. The boat sailed 11 miles in the afternoon. How many miles did the boat end up sailing on that day?

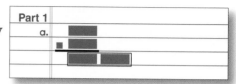

Part 2 — Find the area and perimeter of each figure.

a. 12 in. 8 in. 5 in.

b. 16 cm 7 cm

c. 15 yd 15 yd 14 yd

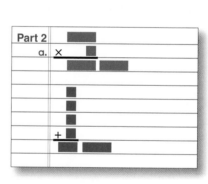

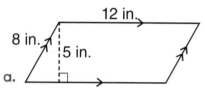

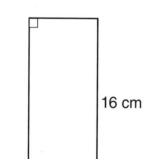

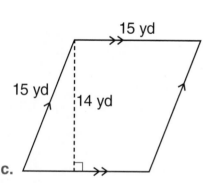

Part 3 — Write the estimation problem for each item.

a. $38\overline{)251}$ b. $47\overline{)343}$ c. $84\overline{)518}$

d. $91\overline{)647}$ e. $57\overline{)135}$ f. $72\overline{)454}$

Lesson 48

Part 4 Copy each problem. Replace one of the letters with a number. Figure out the other letter.

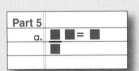

a. $\dfrac{4}{5}H = P$ $\boxed{H = 15}$ b. $\dfrac{3}{2}t = y$ $\boxed{t = 20}$ c. $\dfrac{3}{8}r = m$ $\boxed{m = 9}$

Independent Work

Part 5 Write the letter equation for each sentence.

a. $\dfrac{4}{5}$ of the birds are chirping.

b. $\dfrac{12}{19}$ of the newspapers were delivered.

c. $\dfrac{1}{8}$ of the men wore hats.

d. $\dfrac{7}{10}$ of the cats are eating.

Part 6 Copy and work each problem.

a. $\begin{array}{r} 2.09 \\ \times\ .03 \\ \hline \end{array}$ b. $\begin{array}{r} 3.71 \\ \times\ 3.9 \\ \hline \end{array}$ c. $\begin{array}{r} 482 \\ \times\ .75 \\ \hline \end{array}$

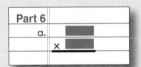

Part 7 Write each fraction as a division problem. Figure out the whole number or mixed number it equals.

a. $\dfrac{703}{5}$ b. $\dfrac{292}{3}$ c. $\dfrac{706}{8}$ d. $\dfrac{437}{7}$

Part 8 Copy and work each problem. If the answer shown is not correct, work the problem again with the correct answer.

a. $34\overline{)203}$ with 6 b. $34\overline{)248}$ with 8 c. $28\overline{)174}$ with 5

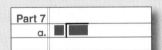

Lesson 49

Part 1 Work each problem.

a. The boat had 146 fish on it. Then the boat unloaded some fish. The boat ended up with 39 fish. How many fish did the boat unload?

b. There were 782 fish on the red boat. There were 311 more fish on the green boat than on the red boat. How many fish were on the green boat?

c. There were 12 rainy days in the fall and 56 rainy days in the spring. How many more rainy days were there in the spring than in the fall?

d. Tom had some money in the morning. He spent $56 and ended up with $18. How much money did he have in the morning?

Part 2 Write the estimation problem for each item.

a. $81\overline{\smash{)}567}$ b. $49\overline{\smash{)}430}$ c. $76\overline{\smash{)}394}$

d. $50\overline{\smash{)}232}$ e. $94\overline{\smash{)}675}$

Part 3 Work each problem.

a. $\frac{2}{3}$ of the fish were perch. There were 48 perch. How many fish were there in all?

b. $\frac{4}{5}$ of the books are library books. There are 24 library books. How many books are there?

c. In a park, $\frac{3}{4}$ of the birds were ducks. There were 15 ducks. How many birds were there?

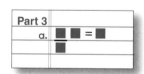

Independent Work

Part 4 Round each number to the nearest thousand.

a. 12,490   b. 9645   c. 2517

Lesson 49

Part 5 Work each problem. Write the answer as a whole number or mixed number and a unit name.

a. The truck started out with $\frac{9}{4}$ tons of gravel. The truck ended up with $\frac{6}{5}$ tons of gravel. How much gravel did the truck drop off?

b. The paper towels lasted for $\frac{32}{5}$ days. The paper napkins lasted for 12 days. How much longer did the napkins last than the towels?

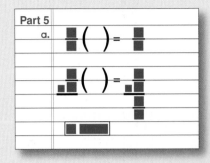

Part 6 Copy and work each problem.

a.   .3 6
   × 4.1

b.   6.0 8
   × 9.2

c.   .2 5
   × .1 7

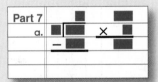

Part 7 Copy and work each problem. If the answer shown is not correct, work the problem again with the correct answer.

a. $62\overline{)419}$ (6)   b. $71\overline{)530}$ (7)   c. $69\overline{)589}$ (9)   d. $75\overline{)641}$ (8)

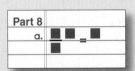

Part 8 Copy each problem. Replace one of the letters with a number. Figure out the other letter.

a. $\frac{5}{8}P = K$ $\boxed{K = 90}$ b. $\frac{3}{2}M = B$ $\boxed{M = 112}$

Part 9 Copy and work each problem. Show the answer as a mixed number.

a. $3\overline{)88}$ b. $9\overline{)60}$ c. $7\overline{)61}$

Lesson 49

Part 10
Write each problem in a column and work it.

a. $\dfrac{5}{4} + \dfrac{3}{6}$ b. $\dfrac{2}{7} + 3$

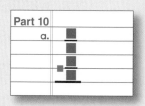

Part 11
Work each problem.

a. At a park there were robins and crows. The ratio of robins to birds was 3 to 4. There were 567 robins. How many crows were there?

b. In a park some of the trees turned red. The ratio of green trees to all trees was 2 to 9. There were 216 trees in the park. How many trees were red?

Part 1 Work each problem.

a. $\frac{2}{5}$ of the dogs had spots. There were 10 dogs with spots. How many dogs were there?

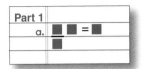

b. $\frac{1}{8}$ of the trees had apples. There were 11 trees with apples. How many trees were there?

c. $\frac{3}{4}$ of the cookies are baked. There are 18 baked cookies. How many cookies are there?

Part 2 Write the number rounded to the arrowed place.

a. 806,572
   ↑

b. 806,572
   ↑

c. 806,572
   ↑

d. 806,572
   ↑

Part 2	
a.	

Part 3 Work each problem.

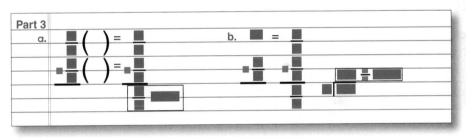

a. The Brown family ate $\frac{10}{8}$ pizzas. The Jackson family ate $\frac{6}{4}$ pizzas. How much more pizza did the Jackson family eat than the Brown family ate?

b. The cabbage was $\frac{3}{4}$ pound lighter than the melon. The melon weighed 4 pounds. How much did the cabbage weigh?

c. A wall was $\frac{10}{3}$ feet high. Workers made the wall $\frac{2}{3}$ feet higher. How high was the wall when they were done?

d. The baker started out with some flour in a bowl. She added $\frac{5}{2}$ cups of flour and ended up with $\frac{7}{2}$ cups of flour in the bowl. How much flour did she start with?

Lesson 50

\boxed{\text{Independent Work}}

Part 4 Find the area and perimeter of each figure.

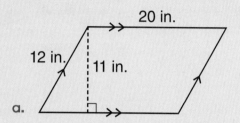

20 in.

12 in. 11 in.

a.

15 ft

9 ft

b.

Part 4	
a.	

Part 5 Copy each problem. Replace one of the letters with a number. Figure out the other letter.

a. $\frac{3}{2} R = C$ $\boxed{C = 792}$ b. $\frac{7}{4} P = T$ $\boxed{T = 560}$

Part 5	
a.	■ ■ _ ■
	■

Part 6 Copy each problem in a column and work it.

a. 53 − 12.24   b. 9 − 2.71   c. 7.3 − 5.06

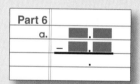

Part 6	
a.	■ . ■
	− ■ . ■
	.

Part 7 Copy and work each problem. If the answer shown is not correct, work the problem again with the correct answer.

a. $89\overline{)624}$ with 6 above

b. $39\overline{)236}$ with 6 above

c. $51\overline{)198}$ with 4 above

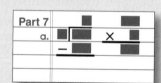

Part 7	
a.	■ ■ × ■
	− ■ ■

Part 8 Copy and work each problem.

a. 1.0 3
   × 6.7

b. .0 4 8
   × .2 6

c. 3 1 8
   × .0 0 9

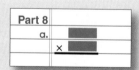

Part 8	
a.	■
	× ■

Lesson 51

Part 1

Find the area of each figure. $A_\triangle = \frac{1}{2}(b \times h)$.

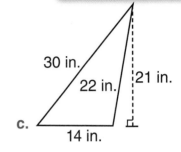

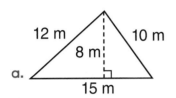

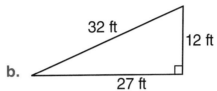

a. (triangle) 12 m, 10 m, 8 m, 15 m

b. (triangle) 32 ft, 12 ft, 27 ft

c. (triangle) 30 in., 22 in., 21 in., 14 in.

Part 1
a. $A_\triangle = \frac{1}{2}(b \times h)$
$A_\triangle = \frac{1}{2}(\blacksquare \times \blacksquare)$
$A_\triangle = \blacksquare$

Part 2

Write each number rounded to the arrowed place.

a. 192,750   b. 192,750   c. 192,750   d. 485,123
   ↑   ↑   ↑   ↑

Part 2
a. $\blacksquare$

Part 3

Work each problem.

a. $\frac{2}{3}$ of the pumpkins were ripe. There were 48 ripe pumpkins. How many pumpkins were there in all?

b. $\frac{5}{9}$ of the children wore jeans. 60 children wore jeans. How many children were there?

c. $\frac{3}{7}$ of the rabbits were white. There were 42 white rabbits. How many total rabbits were there?

d. $\frac{9}{10}$ of the blankets were warm. There were 180 warm blankets. How many blankets were there in all?

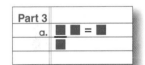

Part 3
a. $\blacksquare$ $\blacksquare$ = $\blacksquare$
$\blacksquare$

Independent Work

Part 4

Write each problem in a column and work it.

a. $46.9 - 11.68$ b. $19.68 + 3.5$ c. $15 - 3.41$

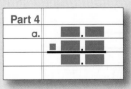

Part 4
a.

Lesson 51

Part 5

Copy and work each problem. If the answer shown is not correct, work the problem again with the correct answer.

a. $65\overline{)484}$ with 8 above

b. $36\overline{)193}$ with 4 above

c. $88\overline{)443}$ with 4 above

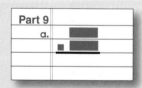

Part 6

Work each problem.

a. In a large building there were 270 doors. The ratio of open doors to all doors was 3 to 9. How many doors were closed?

b. Alex collected stamps and coins. He had 2 stamps for every 9 coins. He had 106 stamps. How many coins were in his collection?

Part 7

Round each number to the nearest thousand.

a. 9820   b. 32,209   c. 8516

Part 8

Copy each problem. Replace one of the letters with a number. Figure out the other letter.

a. $\frac{3}{7}R = T$ [R = 36]

b. $\frac{8}{3}K = B$ [B = 336]

Part 9

Copy and work each problem.

a. $\begin{array}{r} 75.0 \\ \times\ .09 \\ \hline \end{array}$

b. $\begin{array}{r} 38.5 \\ \times\ 9.3 \\ \hline \end{array}$

c. $\begin{array}{r} .270 \\ \times\ .36 \\ \hline \end{array}$

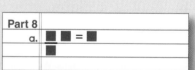

Lesson 51

Part 10 Find the area and perimeter of each figure.

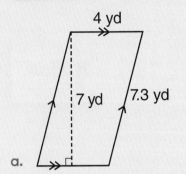

a.

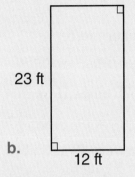

b.

Part 10	
a.	

Part 11 Copy and work each problem. Show the answer as a mixed number.

a. $7\overline{)46}$ b. $5\overline{)38}$ c. $9\overline{)88}$

Part 11	
a.	

Lesson 52

Part 1   Work each problem.

a. $\frac{9}{10}$ of the worms were underground. There were 20 worms. How many underground worms were there?

b. $\frac{3}{5}$ of the birds had nests. There were 25 birds. How many nesting birds were there?

c. $\frac{1}{4}$ of the ice cubes were frozen. There were 48 ice cubes. How many were frozen?

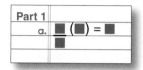

Part 2   Copy each problem. Work an estimation problem for the tens. Then work the original problem.

a. $87\overline{)368}$ b. $50\overline{)426}$ c. $32\overline{)194}$

Part 3   Find the area of each figure. $A_\triangle = \frac{1}{2}(b \times h)$.

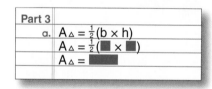

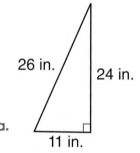

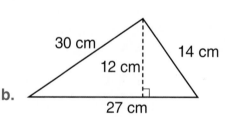

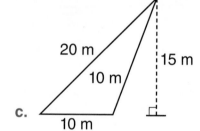

Independent Work

Part 4 Copy each problem. Replace one of the letters with a number. Figure out the
other letter.

a. $\frac{2}{5}T = M$ $\boxed{M = 18}$ b. $\frac{7}{9}F = A$ $\boxed{A = 21}$

c. $\frac{3}{4}R = E$ $\boxed{R = 20}$

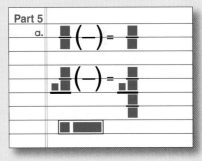

Part 5 Work each problem. Write the answer as a whole number or mixed number and a
unit name.

a. The clock was $\frac{34}{3}$ years older than the rug. The rug was
$\frac{19}{2}$ years old. How old was the clock?

b. After the mail carrier delivered $\frac{28}{3}$ pounds of mail, she still
had $\frac{13}{2}$ pounds of mail to deliver. How much mail did she
start with?

Part 6 Copy and work each problem. If the answer shown is not correct, work the
problem again with the correct answer.

a. $83\overline{\smash)714}^{\,8}$ b. $34\overline{\smash)264}^{\,8}$

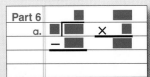

Part 7 Copy each problem in a column and work it.

a. $71.3 - 12.25$ b. $4 - 2.19$ c. $27 - 6.48$

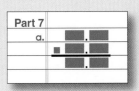

Lesson 52

Part 8 Find the area and perimeter of each figure.

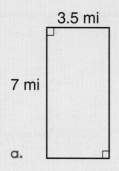

3.5 mi

7 mi

a.

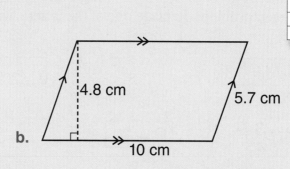

4.8 cm

5.7 cm

b.

10 cm

Part 8	
a.	

Part 9 Work each problem.

a. On a lot there were cars and trucks. There were 4 cars for every 3 trucks. There were 140 vehicles on the lot. How many were cars?

b. The ratio of ripe cherries to unripe cherries on a tree was 5 to 4. There were 480 unripe cherries on the tree. How many cherries were on the tree?

Part 9	
a.	

Part 10 Copy and work each problem. Show the answer as a whole number or a mixed number.

a. $3\overline{\smash{)}45}$ b. $4\overline{\smash{)}73}$ c. $6\overline{\smash{)}29}$

Part 10	
a.	■ ■

Part 11 Work each problem.

a. $\frac{2}{3}$ of the workers were hungry. There were 48 hungry workers. How many workers were there in all?

b. $\frac{3}{8}$ of the cars were dirty. There were 96 dirty cars. How many cars were there in all?

Part 11	
a.	■ ■ = ■
	■

Connecting Math Concepts

Lesson 53

Find the area and perimeter for each figure.

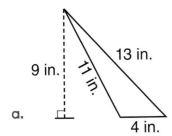

a.

13 in.
9 in.
11 in.
4 in.

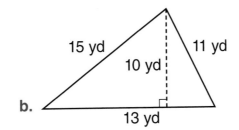

b.

15 yd
10 yd
11 yd
13 yd

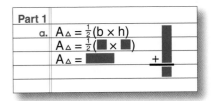

Part 1
a. $A_\triangle = \frac{1}{2}(b \times h)$
$A_\triangle = \frac{1}{2}(\blacksquare \times \blacksquare)$
$A_\triangle = \blacksquare$ +

Part 2 Figure out the prime factors for each number.

a. 30   b. 35   c. 44   d. 32

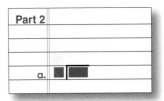

Part 2

a. ■ ■

Part 3 Work each problem.

a. $\frac{2}{3}$ of the rabbits were white. There were 30 rabbits. How many white rabbits were there?

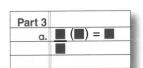

Part 3
a. ■ (■) = ■
■

b. $\frac{7}{10}$ of the trees were planted. There were 49 planted trees. How many trees were there in all?

c. $\frac{2}{5}$ of the camels were sleeping. 56 camels were sleeping. How many camels were there in all?

d. $\frac{3}{8}$ of the cups were empty. There were 160 cups. How many empty cups were there?

Independent Work

Part 4 Copy each problem. Work an estimation problem for the tens. Then work the original problem.

a. $64\overline{)254}$ b. $29\overline{)124}$ c. $34\overline{)176}$

Part 4
a. ■ ■ ■ ■

Lesson 53

Part 5 Work each problem.

a. What's $\frac{3}{5}$ of 80?   b. What's $\frac{8}{7}$ of 45?   c. What's $\frac{2}{3}$ of 180?

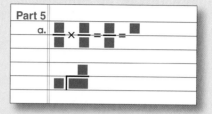

Part 6 Write each problem with the decimal points lined up and work it.

a. 3.07 − 2.5   b. 18.6 − 3.51   c. 82.9 − 6.4

d. 7.61 − .482   e. 46.2 − 1.07

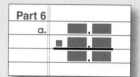

Part 7 Work each problem.

a. At a building site, there were piles of sand and gravel. There were 6 tons of gravel for every 5 tons of sand. There were 140 tons of sand. How many tons of gravel were there?

b. The kennel had dogs and cats. The ratio of dogs to cats was 5 to 4. There were 360 pets. How many cats were there?

Part 8 Copy each problem. Replace one of the letters with a number. Figure out the other letter.

a. $\frac{7}{8}P = V$ $\boxed{V = 28}$ b. $\frac{4}{3}J = T$ $\boxed{J = 27}$

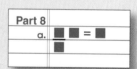

Part 9 Figure out the whole number or mixed number each fraction equals.

a. $\frac{29}{3}$ b. $\frac{86}{4}$ c. $\frac{609}{3}$ d. $\frac{99}{5}$

Lesson

Part 10 Work each problem. Write each answer as a whole number or mixed number with a unit name.

a. Harry ran $\frac{4}{3}$ of a mile on Monday and $\frac{13}{2}$ miles on Friday. How much farther did he run on Friday than on Monday?

Part 10	
a.	

b. The blue boat was $\frac{29}{4}$ meters long. The red boat was $\frac{7}{4}$ meters longer than the blue boat. How long was the red boat?

c. The painters painted $\frac{10}{3}$ houses this week and $\frac{14}{9}$ houses last week. How many houses did they paint during both weeks?

Lesson 54

Part 1
Figure out the prime factors for each number.

    a. 70   **b.** 66   **c.** 34

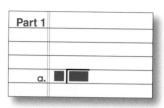

Part 2
Answer both questions.

a. The ratio of big spoons to little spoons was 1 to 4. There were 20 big spoons. How many little spoons were there? How many spoons were there in all?

Part 3
Find the area and perimeter of each figure.

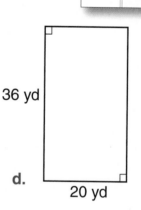

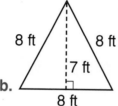

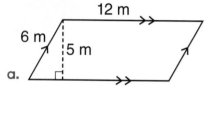

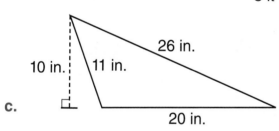

a. 12 m, 6 m, 5 m

b. 8 ft, 8 ft, 7 ft, 8 ft

c. 10 in., 11 in., 26 in., 20 in.

d. 36 yd, 20 yd

Independent Work

Part 4
Copy each problem. Write the estimation problem. Work the problem.

  a. $53\overline{)211}$ **b.** $62\overline{)303}$ **c.** $76\overline{)558}$

Lesson 54

Part 5 Work each problem.

a. $\frac{3}{8}$ of the shirts were wet. 51 shirts were wet. How many shirts were there in all?

b. $\frac{7}{8}$ of the windows were clean. There were 672 windows. How many were clean?

c. $\frac{5}{9}$ of the burgers were cooked. 60 burgers were cooked. How many burgers were there all together?

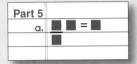

Part 6 Write the place value of each arrowed digit.

a. 378,200   b. 378,200   c. 54,926

Part 7 Find the area and perimeter of each figure.

a. 14 cm, 21 cm, 18 cm

b. 60 cm, 35 cm

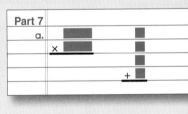

Part 8 Figure out the prime factors for each number.

a. 39   b. 18   c. 25   d. 44   e. 69

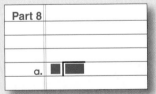

Part 9 Copy and work each problem.

a.
$$\begin{array}{r} .16 \\ \times\ .09 \\ \hline \end{array}$$

b.
$$\begin{array}{r} 4.47 \\ \times\ .13 \\ \hline \end{array}$$

c.
$$\begin{array}{r} .203 \\ \times\ .006 \\ \hline \end{array}$$

Lesson 55

Part 1 Work each ratio problem. Add or subtract to answer the second question.

a. There were white rocks and brown rocks. The ratio of brown rocks to all rocks was 3 to 5. There were 30 rocks. How many brown rocks were there? How many white rocks were there?

b. There were 6 white butterflies for every 2 yellow butterflies. There were 20 yellow butterflies. How many white butterflies were there? How many butterflies were there in all?

Part 2 Copy each problem. Write the estimation problem. Work the problem.

a. $26\overline{)249}$ b. $33\overline{)190}$ c. $55\overline{)333}$

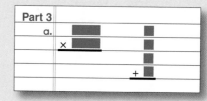

Independent Work

Part 3 Find the area and perimeter of each figure.

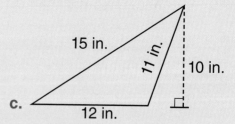

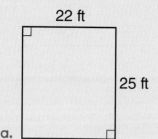

22 ft

25 ft

a.

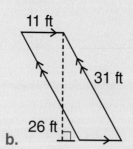

11 ft

31 ft

26 ft

b.

15 in.   11 in.   10 in.

12 in.

c.

Part 4 Copy and work each problem. Use an estimation problem.

a. $53\overline{)179}$ b. $68\overline{)148}$

Lesson 55

Part 5 Write the value of each arrowed digit.

a. 125,051   b. 125,621   c. 348,209

Part 5	
a.	

Part 6 Work each problem.

a. $\frac{1}{8}$ of the workers had the flu. There were 13 sick workers. How many total workers were there?

b. $\frac{5}{7}$ of the butterflies were yellow. There were 200 yellow butterflies. How many butterflies were there in all?

c. $\frac{2}{5}$ of the cans were empty. There were 360 cans. How many empty cans were there?

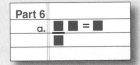

Part 7 Copy each problem. Replace one of the letters with a number. Figure out the other letter.

a. $\frac{8}{3}K = T$ ☐ T = 16 ☐   b. $\frac{2}{9}P = C$ ☐ P = 6 ☐

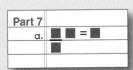

Part 8 Write each problem with the decimal points lined up and work it.

a. 46 − 9.48   b. 12.2 − 8.05   c. 23.9 − 16.9

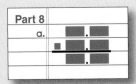

Part 9 Copy and work each problem. Show the answer as a mixed number.

a. 5⟌77   b. 2⟌79   c. 9⟌49

Lesson 56

Part 1
Copy each number. Below, write the rounded number.

a. Round 615,380 to the ten-thousands place.

b. Round 206,485 to the ten-thousands place.

c. Round 506,370 to the thousands place.

d. Round 342,781 to the ten-thousands place.

e. Round 318,960 to the thousands place.

Part 2
Work each problem.

a. There were 9 living trees for every dead tree. There were 72 living trees. How many dead trees were there? How many trees were there in all?

b. The ratio of cats to all pets was 2 to 5. There were 90 pets. How many were cats? How many were not cats?

Independent Work

Part 3
Copy each problem. Write the estimation problem. Work the problem.

a. $76\overline{)624}$ b. $42\overline{)123}$

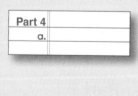

Part 4
Find the area and perimeter of each figure.

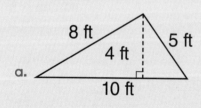

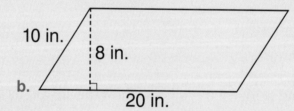

Part 5
Work each problem. Write the answer as a whole number.

a. What's $\frac{7}{8}$ of 112? b. What's $\frac{4}{5}$ of 70? c. What's $\frac{3}{2}$ of 96?

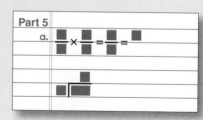

Connecting Math Concepts

Lesson 56

Part 6 Figure out the whole number or mixed number each fraction equals.

a. $\dfrac{246}{3}$ b. $\dfrac{118}{4}$ c. $\dfrac{306}{5}$ d. $\dfrac{490}{7}$

Part 6	
a.	■ ▆

Part 7 Write the place value of each arrowed digit.

a. 384,650 b. 384,650 c. 235,729
   ↑   ↑   ↑

Part 7	
a.	▆

Part 8 Work each problem. Write each answer as a whole number or mixed number with a unit name.

a. Jan read $\dfrac{13}{5}$ more books than Andy read. Jan read 12 books. How many books did Andy read?

Part 8	
a.	

b. The crew dug up $\dfrac{20}{3}$ tons of gravel in the morning and $\dfrac{7}{2}$ tons in the afternoon. How much more gravel did they dig up in the morning than they dug up in the afternoon?

c. Vern read $\dfrac{23}{4}$ books. José read $\dfrac{13}{2}$ books. How many books did both boys read?

Part 9 Copy and work each problem. Use an estimation problem.

a. $77\overline{)432}$ b. $42\overline{)89}$

Part 9	
a.	

Part 10 Copy and work each problem.

a. $\begin{array}{r} 3.24 \\ \times\ .005 \\ \hline \end{array}$ b. $\begin{array}{r} 7.16 \\ \times\ .03 \\ \hline \end{array}$ c. $\begin{array}{r} .203 \\ \times\ .004 \\ \hline \end{array}$

Part 10	
a.	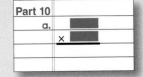

Part 11 Find all the prime factors of each number.

a. 50   b. 54   c. 33   d. 36

Part 11	
a.	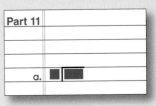

Lesson 57

Part 1 Simplify each fraction.

a. $\dfrac{20}{28}$ b. $\dfrac{6}{30}$ c. $\dfrac{12}{20}$

Part 2 Copy each number. Below, write the rounded number.

a. Round 134,965 to the ten-thousands place.

b. Round 856,390 to the ten-thousands place.

c. Round 204,586 to the ten-thousands place.

d. Round 138,634 to the thousands place.

e. Round 537,821 to the hundreds place.

Part 3 Figure out each blue angle.

a.

122°

b.

18°

c.

235°

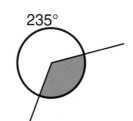

Independent Work

Part 4 Copy each problem. Write the estimation problem. Work the problem.

a. $84\overline{)412}$ b. $37\overline{)349}$

Part 5 Copy and work each problem. Show the answer as a mixed number.

a. $8\overline{)59}$ b. $7\overline{)66}$ c. $7\overline{)53}$

Lesson 57

Part 6 Work each problem.

a. $\frac{3}{4}$ of the workers used hammers. There were 288 workers. How many used hammers?

b. $\frac{9}{10}$ of the rocks were brown. There were 216 brown rocks. How many rocks were there in all?

c. $\frac{4}{5}$ of the goats were sleeping. There were 220 goats. How many were sleeping?

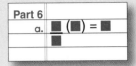

Part 6	
a.	■ (■) = ■
	■

Part 7 Copy and work each problem.

a.
$$\begin{array}{r} 496 \\ \times\ \ \ 8 \\ \hline \end{array}$$

b.
$$\begin{array}{r} 703 \\ \times\ \ 28 \\ \hline \end{array}$$

c.
$$\begin{array}{r} 405 \\ \times\ \ 90 \\ \hline \end{array}$$

Part 7	
a.	▬
	× ▬
	▬

Part 8 Write the place value of each arrowed digit.

a. 219,356
   ↑

b. 219,356
   ↑

c. 219,856
   ↑

Part 8	
a.	▬

Part 9 Copy and work each problem. Use an estimation problem.

a. 32 ⟌ 1 6 6

b. 48 ⟌ 3 5 2

Part 9	
a.	◼▬ ◼▬

Lesson 58

Part 1
Copy each number. Below, write the rounded number.

a. Round 325,806 to the thousands place.

b. Round 175,360 to the ten-thousands place.

c. Round 489,368 to the hundreds place.

d. Round 534,603 to the ten-thousands place.

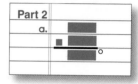

Part 2
Figure out each blue angle.

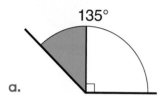

a.

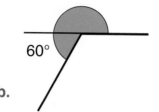

b.

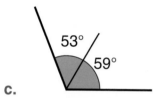

c.

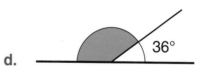

d.

Part 3
Work each problem.

a. There were 7 green bugs for every 3 red bugs in a nest. There were 300 bugs in the nest. How many were green? How many were red?

b. The ratio of men to all adults was 5 to 8. There were 50 men. How many women were there? How many adults were there?

Independent Work

Part 4
Copy and work each problem. Use an estimation problem.

a. $32\overline{)262}$ b. $63\overline{)409}$ c. $57\overline{)471}$

Lesson

Part 5 Find all the prime factors for each number.

    a. 52   b. 27   c. 64   d. 48

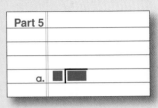

Part 6 Answer each question.

a. How many degrees are in a straight line?

b. How many degrees are in the corner of a rectangle?

c. How many degrees are in a circle?

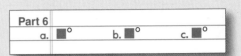

Part 7 Copy and simplify each fraction.

    a. $\dfrac{9}{27}$ b. $\dfrac{16}{20}$ c. $\dfrac{100}{15}$ d. $\dfrac{12}{28}$

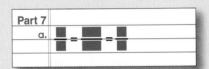

Part 8 Write each problem with the decimal points lined up and work it.

a. $14.03 - 7.28$ b. $12.36 - 10.4$ c. $7.56 + 11.082$

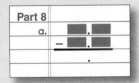

Part 9 Copy and work each division fact.

a. $9\overline{)8\,1}$ b. $7\overline{)4\,2}$ c. $8\overline{)5\,6}$ d. $8\overline{)7\,2}$

e. $3\overline{)2\,1}$ f. $9\overline{)5\,4}$ g. $5\overline{)4\,5}$ h. $7\overline{)6\,3}$

Lesson 58

Part 10

Find the area and perimeter of each figure.

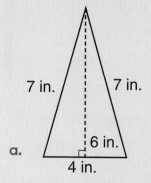

a. (triangle) 7 in., 7 in., 6 in., 4 in.

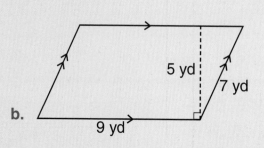

b. (parallelogram) 5 yd, 7 yd, 9 yd

Part 10	
a.	

Part 11

Copy and work each problem.

a.
$$\begin{array}{r} 8.97 \\ \times\ .34 \\ \hline \end{array}$$

b.
$$\begin{array}{r} .606 \\ \times\ .04 \\ \hline \end{array}$$

c.
$$\begin{array}{r} 2.20 \\ \times\ .38 \\ \hline \end{array}$$

Part 11		
a.		
	×	

Lesson 59

Part 1 Write each rounded number.

a. Round 8,463,582 to the hundred-thousands place.

b. Round 23,430 to the hundreds place.

c. Round 4,568,200 to the ten-thousands place.

d. Round 10,793 to the thousands place.

e. Round 542,383 to the thousands place.

Part 1	
a.	▬

Part 2 Copy each problem. Rewrite below and simplify. Then write the answer.

a. $\dfrac{11}{8} \times \dfrac{56}{1}$ b. $40 \times \dfrac{6}{5}$ c. $\dfrac{81}{1} \times \dfrac{5}{9}$ d. $\dfrac{3}{4} \times 20$

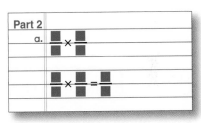

Part 3 Figure out each blue angle.

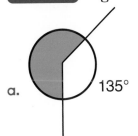

a.   135°

b.   120°

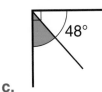

c.   48°

d.   100°

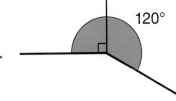

Part 3	
a.	

Part 4 Work each problem. Answer both questions.

a. There were trees and bushes in the woods. There were 5 trees for every 7 plants. There were 48 bushes. How many plants were there? How many trees were there?

b. The ratio of new books to old books on a shelf was 2 to 9. There were 72 old books on the shelf. How many books were on the shelf? How many new books were on the shelf?

Part 4	
a.	

Lesson 59

Independent Work

Part 5 Find the area and perimeter of each figure.

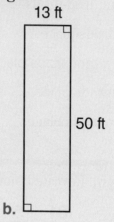

13 ft

50 ft

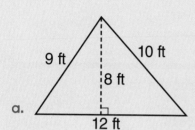

9 ft
10 ft
8 ft

a.

12 ft

b.

Part 5	
a.	

Part 6 Work each problem.

a. $\frac{1}{9}$ of the apples were rotten. There were 63 rotten apples. How many apples were there in all?

b. $\frac{5}{6}$ of the machines were running. There were 120 machines. How many were running?

c. $\frac{2}{3}$ of the students were on the playground. There were 150 students on the playground. How many students were there in all?

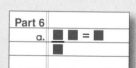

Part 6	
a.	■ ■ = ■
	■

Part 7 Figure out the whole number or mixed number each fraction equals.

a. $\frac{46}{8}$ b. $\frac{50}{7}$ c. $\frac{38}{5}$ d. $\frac{72}{8}$

Part 7	
a.	■ ■

Part 8 Find all the prime factors for each number.

a. 24   b. 46   c. 35   d. 63

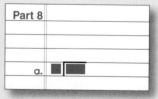

Part 8	
a.	■ ■

Connecting Math Concepts

Lesson  59

Part 9 Copy and work each problem. Use an estimation problem.

a. $53\overline{)273}$ b. $48\overline{)239}$ c. $79\overline{)490}$

Part 10 Write each problem with the decimal points lined up and work it.

a. $3.17 - 1.8$ b. $13.06 - 2.59$ c. $11 + 6.50$

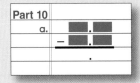

Part 11 Copy and simplify each fraction.

a. $\dfrac{6}{14}$ b. $\dfrac{63}{36}$ c. $\dfrac{48}{16}$

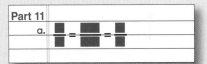

Lesson 60

Part 1
Write the base and exponent for each item.

a. $9 \times 9 \times 9 \times 9$ b. $20 \times 20 \times 20 \times 20 \times 20$ c. $13 \times 13 \times 13$

d. 6×6 e. $8 \times 8 \times 8 \times 8 \times 8$

Part 2
Write each rounded number.

a. Round 3,257,835 to the hundred-thousands place.

b. Round 17,804 to the thousands place.

c. Round 3449 to the hundreds place.

d. Round 76,020 to the ten-thousands place.

e. Round 9,312,580 to the hundred-thousands place.

Part 3
Work each problem. Answer both questions.

a. The ratio of girls to children in the park was 6 to 11. There were 30 girls. How many boys were there? How many children were there?

b. There were 10 fleas for every 3 mice. There were 60 fleas. How many animals were there? How many mice were there?

Part 4
Figure out each blue angle.

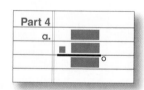

a.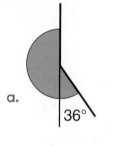
36°

b.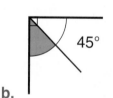
45°

c.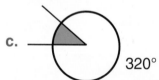
320°

d.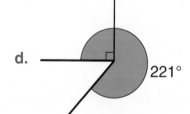
221°

Lesson 60

Independent Work

Part 5 Answer each question.

a. How many degrees are in the corner of a rectangle?

b. How many degrees are in a circle?

c. How many degrees are in a straight line?

Part 5 | a. ■° | b. ■° | c. ■°

Part 6 Copy and work each division fact.

a. $9\overline{)63}$ b. $4\overline{)36}$ c. $8\overline{)56}$ d. $8\overline{)64}$

e. $3\overline{)27}$ f. $7\overline{)63}$ g. $9\overline{)54}$ h. $9\overline{)72}$

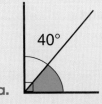

Part 6 | a.

Part 7 Copy and work each problem.

a. $\begin{array}{r} 4.7 \\ \times\ 8.6 \\ \hline \end{array}$ b. $\begin{array}{r} .179 \\ \times\ .04 \\ \hline \end{array}$

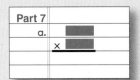

Part 7 | a. ×

Part 8 Figure out the shaded angle.

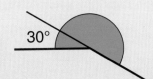

a. 40°

b. 30°

Part 8 | a. ■ °

Part 9 Copy and work each problem. Use an estimation problem.

a. $52\overline{)375}$ b. $67\overline{)348}$ c. $73\overline{)592}$

Part 9 | a.

Part 10 Figure out the whole number or mixed number each fraction equals.

a. $\dfrac{49}{8}$ b. $\dfrac{63}{9}$ c. $\dfrac{30}{4}$ d. $\dfrac{50}{6}$

Part 10 | a.

Lesson 61

Part 1 Copy each problem. Simplify then multiply.

a. $\dfrac{7}{5} \times \dfrac{40}{35}$ b. $\dfrac{3}{5} \times \dfrac{40}{27}$ c. $\dfrac{15}{8} \times \dfrac{64}{3}$ d. $\dfrac{3}{28} \times \dfrac{4}{21}$

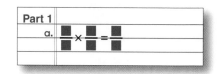

Part 2 Write the base and exponent for each item.

a. $12 \times 12 \times 12 \times 12$

b. $2 \times 2 \times 2$

c. 8×8

d. $5 \times 5 \times 5 \times 5 \times 5$

Part 3 Work each problem.

a. The ratio of hens to roosters was 9 to 2. There were 110 birds. How many roosters were there? How many hens were there?

b. On the beach walk, there were 3 broken shells for every 5 shells we found. We found 16 unbroken shells. How many broken shells did we find? How many shells did we find in all?

Independent Work

Part 4 Figure out each blue angle.

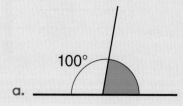

a.

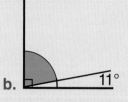

b.

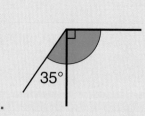

c.

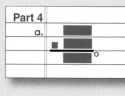

Part 5 Write each problem with the decimal points lined up and work it.

a. $1.2 + 17.59$ b. $24.06 - 3.88$ c. $7 - 5.74$

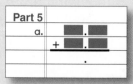

Lesson 61

Part 6 Figure out the whole number or mixed number each fraction equals.

a. $\dfrac{55}{9}$ b. $\dfrac{56}{8}$ c. $\dfrac{32}{7}$ d. $\dfrac{60}{7}$

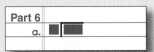

Part 7 Copy and work each division fact.

a. $7\overline{)56}$ b. $6\overline{)48}$ c. $9\overline{)72}$ d. $9\overline{)54}$

e. $6\overline{)42}$ f. $6\overline{)36}$ g. $9\overline{)63}$ h. $8\overline{)64}$

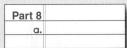

Part 8 Find the area and perimeter of each figure.

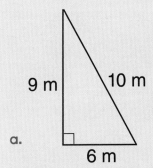

a.

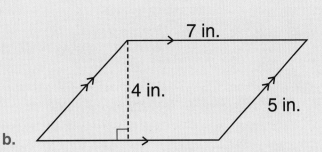

b.

Part 9 Work each problem.

a. $\dfrac{7}{10}$ of the tools were wrenches. There were 98 wrenches. How many tools were there?

b. $\dfrac{2}{7}$ of the batteries were weak. There were 84 batteries in all. How many batteries were weak?

c. $\dfrac{4}{7}$ of the students liked math. There were 56 students who liked math. How many students were there?

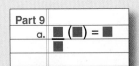

Part 10 Copy and work each problem. Use an estimation problem.

a. $67\overline{)219}$ b. $88\overline{)453}$ c. $73\overline{)396}$

Lesson 61

Part 11 Copy and simplify each fraction.

a. $\dfrac{32}{8}$ b. $\dfrac{56}{49}$ c. $\dfrac{24}{40}$

Part 12 Find all the prime factors for each number.

a. 36   b. 18   c. 46   d. 44

Part 12	
a.	

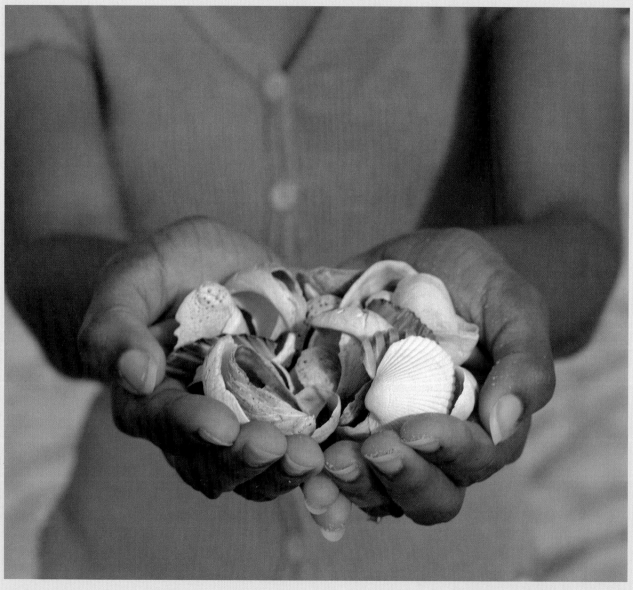

Connecting Math Concepts

Lesson 62

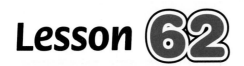

Part 1 Copy each problem. Simplify before you multiply.

a. $\dfrac{21}{4}\left(\dfrac{8}{3}\right) = $ ■   b. $\dfrac{3}{15}\left(\dfrac{2}{20}\right) = $ ■   c. $\dfrac{7}{9}\left(\dfrac{27}{28}\right) = $ ■

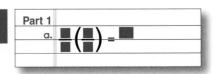

Part 2 Write the base and exponent for each item.

a. 5×5 b. 9 c. $6 \times 6 \times 6$

d. $1 \times 1 \times 1 \times 1$ e. $k \times k \times k \times k \times k$ f. $b \times b$

Part 2	
a.	■■

Independent Work

Part 3 Work each problem.

a. $\dfrac{3}{5}$ of the saws were rusty. There were 45 saws. How many were rusty?

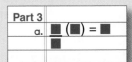

b. $\dfrac{8}{9}$ of the shoes were clean. There were 144 clean shoes. How many shoes were there?

c. $\dfrac{1}{4}$ of the men were resting. 36 men were resting. How many men were there?

Part 4 Write each rounded value.

a. Round 1,685,433 to the ten-thousands place.

b. Round 46,460 to the hundreds place.

c. Round 418,325 to the thousands place.

d. Round 3,206,499 to the ten-thousands place.

Part 4	
a.	■

Lesson 62

Part 5
Figure out the blue angle.

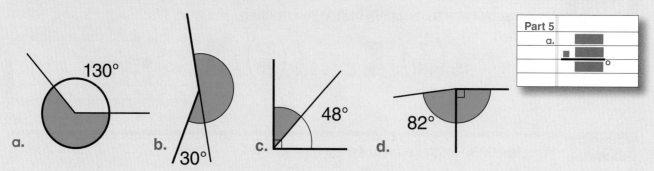

a.   130°

b.   30°

c.   48°

d.   82°

Part 5	
a.	▬▬
	■ ▬▬
	▬▬ °

Part 6
Work each problem. Add or subtract to answer the second question.

a. The ratio of boys to girls at the beach was 6 to 7. There were 210 boys at the beach. How many children were there? How many girls were there?

b. There were 4 sick goats for every 23 goats. There were 69 goats in all. How many were sick? How many were well?

c. There were 7 clean windows for every 8 windows. There were 224 clean windows. How many windows were dirty? How many windows were there in all?

Part 6	
a.	

Part 7
Copy and work each problem. Use an estimation problem.

a. 48⟌204   b. 52⟌386   c. 71⟌325

Part 7	
a.	■▬ ■▬

Part 8
Find the area and perimeter of each figure.

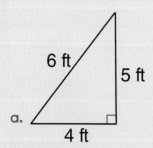

a.   6 ft   5 ft   4 ft

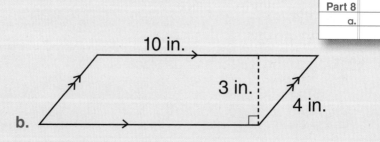

b.   10 in.   3 in.   4 in.

Part 8	
a.	

Lesson

Part 1 Copy each fraction. Figure out the greatest common factor. Write the simplified fraction.

a. $\dfrac{16}{24}$ b. $\dfrac{20}{25}$ c. $\dfrac{45}{18}$ d. $\dfrac{28}{35}$

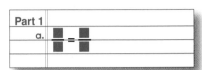

Part 2 Copy and work each problem.

a.
```
  2004
-  165
```
b.
```
  1007
-   59
```
c.
```
  5003
- 1426
```
d.
```
  2005
- 1806
```

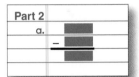

Part 3 Copy each item. Show the multiplication and what it equals.

a. 12^2 b. 6^3 c. 2^5 d. 10^3 e. 3^1

Independent Work

Part 4 Copy each problem. Simplify. Then multiply.

a. $\dfrac{3}{5} \times \dfrac{10}{27}$ b. $\dfrac{7}{4}(32)$ c. $\dfrac{3}{15}\left(\dfrac{2}{4}\right)$

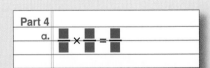

Part 5 Write the rounded number.

a. Round 625,408 to the thousands place.

b. Round 2,512,000 to the millions place.

c. Round 826,193 to the ten-thousands place.

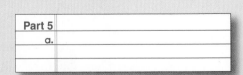

Part 6 Copy and work each problem.

a.
```
   .94
×  7.6
```
b.
```
   .038
×   .14
```

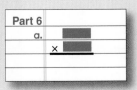

Connecting Math Concepts

Lesson 63

Part 7
Write the answer to each question.

a. How many degrees are in a circle?

b. How many degrees are in the corner of a rectangle?

c. How many degrees are in a straight line?

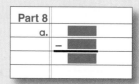

Part 7	
a.	

Part 8
Copy and work each problem.

a.
```
  1 4 0 0
 −  9 2 7
```

b.
```
  8 0 0 6
 − 5 2 4 5
```

c.
```
  2 0.6 0
 −   1.5 3
```

Part 8	
a.	

Part 9
Work each problem. Add or subtract to answer the second question.

a. In the city, the ratio of wet streets to dry streets was 2 to 9. There were 72 dry streets. How many streets were wet? How many streets were there in all?

Part 9	
a.	

b. The ratio of farmers wearing hats to all farmers was 2 to 7. There were 140 farmers. How many farmers were not wearing hats? How many farmers were wearing hats?

Part 10
Use estimation to work each problem.

a. 63⟌268   b. 49⟌325   c. 52⟌381

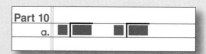

Connecting Math Concepts

Lesson

Part 1 Copy each fraction. Figure out the greatest common factor. Write the simplified fraction.

a. $\dfrac{6}{9}$ b. $\dfrac{30}{24}$ c. $\dfrac{15}{25}$ d. $\dfrac{24}{16}$ e. $\dfrac{16}{20}$

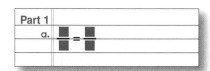

Part 2 Copy and work each problem.

a. 8 0 0 9
  – 1 2 7

b. 8 0 0 0
  – 1 2 7

c. 7 0 0 3
 –1 0 4 5

d. 7 0 0 3
 –1 0 4 2

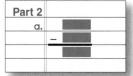

Part 3 Copy each item. Show the multiplication and what it equals.

a. 4^3 b. 2^4 c. 3^2 d. 2^3 e. 5^2

Independent Work

Part 4 Copy and work each problem.

a.   7 8
 × 6.8

b.   .5 9
 × .5

c.   2 8 7
 × 7 9

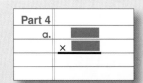

Part 5 Copy and work each problem.

a. $4\overline{)2\,4.8}$ b. $3\overline{)6.1\,8}$ c. $5\overline{)1\,5.2\,0}$

Lesson 64

Part 6 Work each problem.

a. There were 56 students. $\frac{7}{8}$ of the students passed the test. How many students passed the test?

b. $\frac{4}{9}$ of the flowers were blooming. There were 100 flowers that were blooming. How many flowers were there in all?

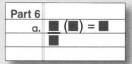

Part 7 Figure out the blue angle.

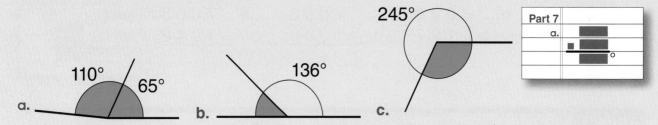

245°

110° 65°

136°

a.

b.

c.

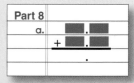

Part 8 Write each problem with the decimal points lined up and work it.

a. $6.3 + 4.53$ b. $12.07 - 5.19$ c. $26.58 - 2.9$

Part 9 Copy each problem. Simplify. Then multiply.

a. $7\left(\frac{3}{49}\right) = \blacksquare$ b. $\frac{8}{10}\left(\frac{5}{16}\right) = \blacksquare$ c. $\frac{4}{9}\left(\frac{27}{16}\right) = \blacksquare$

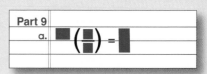

Lesson

Part 1 Copy each fraction. Figure out the greatest common factor. Write the simplified fraction.

a. $\dfrac{40}{20}$ b. $\dfrac{15}{18}$ c. $\dfrac{35}{14}$ d. $\dfrac{54}{63}$ e. $\dfrac{48}{40}$

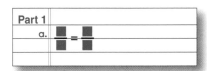

Part 2 Copy and work each problem.

a. $\begin{array}{r} 8000 \\ -\ \ 195 \\ \hline \end{array}$ b. $\begin{array}{r} 4004 \\ -\ 524 \\ \hline \end{array}$ c. $\begin{array}{r} 5029 \\ -\ \ \ 73 \\ \hline \end{array}$ d. $\begin{array}{r} 6403 \\ -1245 \\ \hline \end{array}$ e. $\begin{array}{r} 6005 \\ -\ \ \ 49 \\ \hline \end{array}$

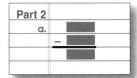

Part 3 Copy each item. Show the multiplication and what it equals.

a. 4^2 b. 1^3 c. 5^3 d. 10^4 e. 2^6

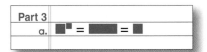

Independent Work

Part 4 Use estimation to work each problem.

a. $76\overline{)449}$ b. $87\overline{)261}$ c. $74\overline{)683}$

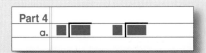

Part 5 Work each problem. Add or subtract to answer the second question.

a. The ratio of trucks to dirty trucks was 5 to 3. There were 40 clean trucks on the lot. How many trucks were dirty? How many trucks were there in all?

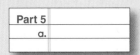

b. There were maple trees and oak trees in a grove. The ratio of maple trees to oak trees was 2 to 9. There were 36 maple trees in the grove. How many trees were in the grove? How many oak trees were in the grove?

Lesson 65

Part 6 — Write the answer to each question.

a. How many degrees are in a circle?

b. How many degrees are in the corner of a rectangle?

c. How many degrees are in a straight line?

Part 6	
a.	

Part 7 — Copy and work each problem.

a. $4\overline{)2.8\,8}$ b. $3\overline{)1.5\,9}$ c. $5\overline{)3.5\,1\,0}$

Part 8 — Write the rounded number.

a. Round 812,773 to the hundreds place.

b. Round 905,196 to the ten-thousands place.

c. Round 1,076,815 to the thousands place.

Part 8	
a.	

Part 9 — Copy and work each problem.

a. 4 3 0 2
  − 6 1 2

b. 8 0 0 5
  − 7 1 6

c. 2 0.7 3
  − .8 8

d. 5 1 0 8
  − 3 2 7

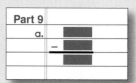

Part 10 — Copy each problem. Simplify. Then multiply.

a. $\dfrac{3}{5}\left(\dfrac{25}{12}\right) = \blacksquare$ b. $\dfrac{1}{6}(42) = \blacksquare$ c. $\dfrac{3}{9}\left(\dfrac{7}{28}\right) = \blacksquare$

Lesson 66

Part 1
Copy each fraction. Figure out the greatest common factor. Write the simplified fraction.

a. $\dfrac{20}{50}$ b. $\dfrac{9}{72}$ c. $\dfrac{9}{12}$ d. $\dfrac{80}{30}$ e. $\dfrac{28}{49}$ f. $\dfrac{18}{60}$

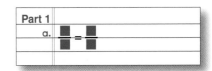

Part 2
Write each number rounded to the arrowed place.

a. 596,014   b. 309,416   c. 469,503   d. 519,293

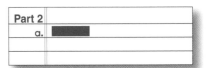

Independent Work

Part 3
Copy and work each problem.

a. $2\overline{)1.8\,6}$ b. $5\overline{)4.0\,1\,5}$ c. $6\overline{)3\,6.2\,4}$

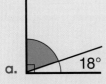

Part 4
Figure out the shaded angle.

a. 18°   b. 78°   c. 80° 95°

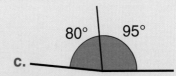

Part 5
Copy and work each problem.

a.
$$\begin{array}{r} 3\,0.8 \\ \times\ \ \ 6.9 \\ \hline \end{array}$$

b.
$$\begin{array}{r} .7\,6\,5 \\ \times\ \ \ .2\,8 \\ \hline \end{array}$$

c.
$$\begin{array}{r} 2\,1\,7 \\ \times\ \ \ 7\,9 \\ \hline \end{array}$$

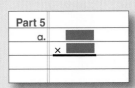

Lesson 66

Part 6 Copy and simplify each fraction. Use the greatest common factor.

a. $\dfrac{48}{16}$　　　　b. $\dfrac{30}{56}$　　　　c. $\dfrac{24}{18}$

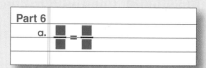

Part 7 Work each problem.

a. $\dfrac{4}{10}$ of the bees were flying. There were 60 bees. How many were flying?

b. $\dfrac{2}{7}$ of the puddles were frozen. There were 30 puddles that were frozen. How many puddles were there in all?

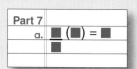

Part 8 Copy each item. Show the repeated multiplication and the answer.

a. 11^3　　　　b. 3^5　　　　c. 2^4

Part 9 Copy and work each problem.

a.
$$\begin{array}{r} 4050 \\ -\ 163 \\ \hline \end{array}$$

b.
$$\begin{array}{r} 3002 \\ -\ 156 \\ \hline \end{array}$$

c.
$$\begin{array}{r} 8001 \\ -\ 700 \\ \hline \end{array}$$

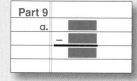

Connecting Math Concepts

Lesson

Part 1 Write each rounded number.

   a. Round .2364 to the tenths place.

   b. Round .2364 to the hundredths place.

   c. Round .2364 to the thousandths place.

Part 2 Write an equation for each item.

   a. 10^4 b. 10^2 c. 10^5 d. 10^3

Part 2	
a.	10■ = ■

Part 3 Write an equation for each item.

a. 1000   b. 1,000,000   c. 10,000   d. 100

Part 3	
a.	■ = 10■

Independent Work

Part 4 Copy and simplify each fraction.

   a. $\dfrac{12}{20}$ b. $\dfrac{90}{66}$ c. $\dfrac{54}{12}$

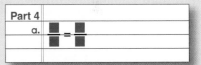

Part 5 Use estimation to work each problem.

a. $86\overline{)265}$ b. $43\overline{)281}$ c. $67\overline{)548}$

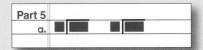

Part 6 Copy and work each problem.

   a. $\begin{array}{r} 2008 \\ -\ 869 \\ \hline \end{array}$ b. $\begin{array}{r} 7006 \\ -6345 \\ \hline \end{array}$ c. $\begin{array}{r} 8004 \\ -3917 \\ \hline \end{array}$

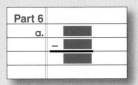

Lesson 67

Part 7 Work each problem. Add or subtract to answer the second question.

a. The ratio of long worms to short worms in a can was 4 to 3. There were 140 worms in the can. How many were short? How many were long?

Part 7	
a.	

b. There were red lizards and green lizards in a zoo. For every 4 lizards, 3 were green. There were 24 red lizards. How many green lizards were there? How many lizards were there in all?

Part 8 Copy each item. Show the repeated multiplication and the answer.

a. 19^2 b. 5^3 c. 3^4

Part 9 Copy and work each problem.

a. $7\overline{).0217}$ b. $2\overline{)2.40}$ c. $4\overline{).0124}$

Part 9	
a.	

Part 10 Write each rounded number.

a. Round 149,762 to the thousands place.

b. Round 214,915 to the ten-thousands place.

Part 10	
a.	

c. Round 138,960 to the hundreds place.

d. Round 4,128,417 to the millions place.

Part 11 Copy each problem. Simplify before you multiply.

a. $\dfrac{3}{8} \times \dfrac{40}{12} =$    b. $\dfrac{28}{5} \left(\dfrac{45}{7} \right) =$ c. $\dfrac{5}{20} \times \dfrac{9}{54} =$

Part 11	
a.	

Lesson 68

Part 1 Write each rounded number.

    a. Round .5874 to the hundredths place.

    b. Round .62718 to the thousandths place.

    c. Round .0127 to the hundredths place.

    d. Round .05281 to the tenths place.

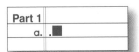

Part 2 Write a $\frac{y}{x}$ equation for each sentence.

    a. The ratio of beans to peas was 9 to 11.

    b. There were 12 phones for every 5 rooms.

    c. Every page had 25 lines.

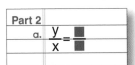

Part 3 Write an equation for each item.

a. 10,000   **b.** 10^5 **c.** 10^3

d. 1,000,000   **e.** 10^1 **f.** 100

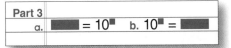

Independent Work

Part 4 Copy and work each problem.

    a. $\begin{array}{r} 451 \\ \times\ \ 76 \\ \hline \end{array}$ **b.** $\begin{array}{r} 832 \\ \times\ \ .4 \\ \hline \end{array}$ **c.** $\begin{array}{r} 187 \\ \times\ \ 36 \\ \hline \end{array}$

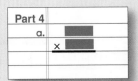

Part 5 Figure out the shaded angle.

a.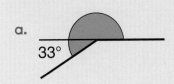

b.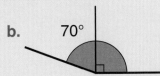

c.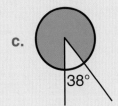

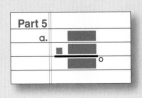

Lesson 68

Part 6 Simplify each fraction.

a. $\dfrac{16}{32}$ b. $\dfrac{25}{45}$ c. $\dfrac{49}{35}$ d. $\dfrac{60}{27}$

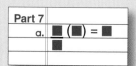

Part 7 Work each problem.

a. $\dfrac{3}{4}$ of the lions were sleeping. There were 36 lions. How many lions were sleeping?

b. $\dfrac{4}{5}$ of the tigers were hungry. 20 tigers were hungry. How many tigers were there?

Part 8 Copy each item. Show the repeated multiplication. Figure out the number it equals.

a. 11^3 b. 3^3 c. 2^5

Part 9 Write each rounded number.

    a. Round 689,512 to the thousands place.

    b. Round 143,260 to the ten-thousands place.

    c. Round 799,720 to the thousands place.

    d. Round 20,530 to the thousands place.

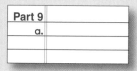

Part 10 Copy and work each problem.

a.	3 0 0 5	b.	7 0 0 3	c.	9 2,0 0 1
	$-$ 9 6		$-$ 3 1 4 3		$-$ 1 0 1 3

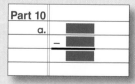

Part 11 Work each problem.

a. There were pigs and sheep on a ranch. The ratio of pigs to sheep was 3 to 9. There were 216 sheep on the farm. How many animals were there? How many pigs were there?

Part 11	
a.	

b. In a field there were 3 brown grasshoppers for every 4 grasshoppers. 48 grasshoppers were not brown. How many grasshoppers were there in all? How many grasshoppers were brown?

Part 12 Copy each problem. Replace one of the letters with a number. Figure out the other letter.

a. $\frac{3}{7} B = C$ $\boxed{C = 120}$

b. $\frac{5}{2} T = M$ $\boxed{T = 50}$

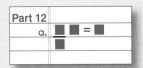

Lesson

Part 1 Write each rounded number.

a. Round .0048 to the hundredths place.

b. Round .0048 to the tenths place.

c. Round .0048 to the thousandths place.

Part 1	
a.	.■■

Independent Work

Part 2 Copy each problem. Replace one of the letters with a number. Figure out the other letter.

a. $\dfrac{2}{3}K = B$ $\boxed{B = 66}$

b. $\dfrac{2}{5}R = J$ $\boxed{R = 10}$

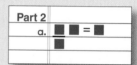

Part 2	
a.	■ ■ = ■ ■

Part 3 Copy each item. Show the repeated multiplication. Figure out the number it equals.

a. 5^3 b. 13^2 c. 9^3

Part 3	
a.	■■ = ■■■ = ■

Part 4 Find the area and perimeter of each figure.

Part 4	
a.	

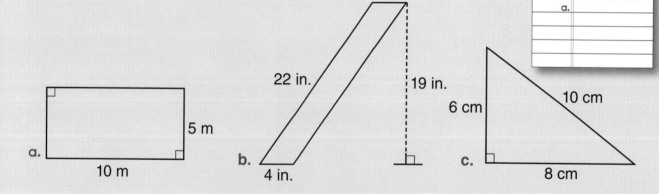

a. (rectangle: 10 m, 5 m)

b. (triangle: 22 in., 4 in., 19 in.)

c. (triangle: 6 cm, 10 cm, 8 cm)

Lesson 69

Part 5 Write each rounded number.

    a. Round 390,713 to the thousands place.

    b. Round 265,320 to the ten-thousands place.

    c. Round 409,380 to the thousands place.

    d. Round 799,651 to the thousands place.

Part 5	
a.	

Part 6 Answer each question.

a. There were 4 sleeping lions for every 5 lions. 60 lions were sleeping. How many lions were there? How many were awake?

Part 6	
a.	

b. The guests had pie or cake for dessert. The ratio of pies to cakes was 3 to 8. There were 44 desserts. How many were pies? How many were cakes?

Part 7 Write the complete equation for each item.

a. $10^3 = $ 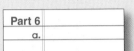   **b.** $1000 = $ ■■   **c.** $100,000 = $ ■■

Part 7	
a.	$10^{■} = $

Part 8 Copy and work each problem.

    a. $\quad 3007$
        $\underline{-\ 166}$

    b. $\quad 8004$
        $\underline{-\ 199}$

    c. $\quad 2007$
        $\underline{-1327}$

Part 8	
a.	■■
	■■
	− ■■

Lesson 70

Part 1 Copy each problem and write the answer.

a. $56 \div 7$ b. $27 \div 9$ c. $45 \div 5$ d. $90 \div 9$

e. $64 \div 8$ f. $36 \div 4$ g. $72 \div 9$ h. $28 \div 4$

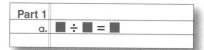

Part 2 Find the volume of each rectangular prism.

a.
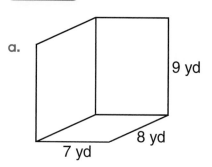
9 yd
8 yd
7 yd

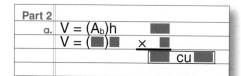

b.
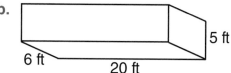
5 ft
6 ft
20 ft

c.

3 mi
13 mi
8 mi

Part 3 Write each rounded number.

a. Round .9876 to the tenths place.

b. Round .0095 to the thousandths place.

c. Round .0962 to the hundredths place.

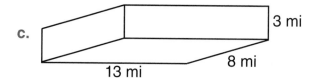

Part 4 Write the equation or the ratio fractions for each sentence.

a. There were 4 black bricks for every 3 red bricks.

b. $\frac{7}{10}$ of the animals were hungry.

c. There were 6 pumpkins for every 11 napkins.

d. There were 5 girls for every 9 children.

e. $\frac{3}{8}$ of the children were girls.

Connecting Math Concepts

Lesson 70

Independent Work

Part 5 Simplify each fraction.

a. $\dfrac{24}{39}$ b. $\dfrac{45}{80}$ c. $\dfrac{26}{14}$

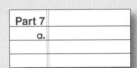

Part 6 Write an equation for each item.

a. 10,000   b. 10^5 c. 1,000,000

Part 6	
a.	▬ = 10■

Part 7 Write each rounded value.

a. Round .3752 to the hundredths place.

b. Round .045 to the tenths place.

c. Round .1392 to the thousands place.

Part 7	
a.	

Part 8 Copy and work each problem.

a. 4006
  −3765

b. 8008
  −4299

c. 3003
  −2012

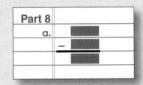

Part 9 Copy each letter equation. Replace one letter with a number. Figure out the other letter.

a. $\dfrac{3}{2}K = P$ K = 126

b. $\dfrac{4}{9}T = J$ J = 72

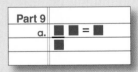

Lesson 70

Part 10 Write each rounded number.

   a. Round 308,601 to the thousands place.

   b. Round 51,832 to the hundreds place.

   c. Round 279,405 to the thousands place.

   d. Round 398,210 to the ten-thousands place.

Part 10	
a.	

Part 11 Work each problem.

a. $\frac{2}{3}$ of the boxes were sealed. There were 120 sealed boxes. How many boxes were there in all?

b. There were 32 spiders. $\frac{5}{8}$ of the spiders were females. How many spiders were female?

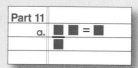

Part 12 Copy each problem. Simplify before you multiply.

a. $\frac{1}{10}\left(\frac{70}{11}\right) = $ ▮ **b.** $45 \times \frac{3}{5} = $ ▮ **c.** $\frac{2}{7}\left(\frac{49}{10}\right) = $ ▮

Part 12	
a.	

Lesson 71

Part 1
Copy each problem and write the answer.

a. 32 ÷ 8   b. 90 ÷ 9   c. 49 ÷ 7   d. 60 ÷ 10

e. 42 ÷ 6   f. 64 ÷ 8   g. 35 ÷ 5   h. 18 ÷ 9

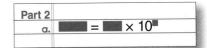

Part 2
Write an equation for each item.

a. 4090   b. 300,200   c. 375,000

d. 190,200   e. 163,000

$$\blacksquare = \blacksquare \times 10^{\blacksquare}$$

Part 3
Find the volume of each rectangular prism.

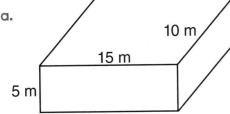

a.

10 m
15 m
5 m

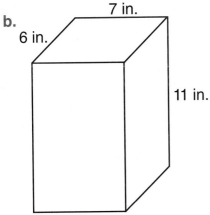

b.

7 in.
6 in.
11 in.

c.

23 cm
5 cm
4 cm

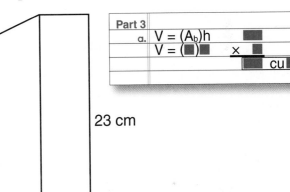

Part 3

a. $V = (A_b)h$ ■
   $V = (\blacksquare)\blacksquare$ × ■
   ■ cu ■

Part 4
Write each rounded value.

a. Round .9842 to the tenths place.

b. Round .0094 to the thousandths place.

c. Round .0991 to the hundredths place.

d. Round .9395 to the tenths place.

Part 4
a. .■

Lesson 71

Part 5
Copy each value. Then round each value to the hundredths.

a. .497   b. .297   c. .697

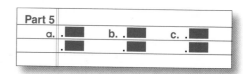

Part 6
For some sentences you'll write a fraction-multiplication equation. For some sentences you'll write a letter ratio and a number ratio.

a. $\frac{9}{10}$ of the bricks were red.

b. There were 7 cats for every 9 squirrels.

c. There were 9 bikes for every 4 motorcycles.

d. $\frac{1}{6}$ of the children were sick.

e. $\frac{6}{11}$ of the books were new.

f. Every 2 boxes weigh 25 pounds.

Part 6	
a.	

Independent Work

Part 7
Copy and work each problem.

a.
$$\begin{array}{r} .86 \\ \times\ 78 \\ \hline \end{array}$$

b.
$$\begin{array}{r} 6964 \\ \times\quad\ 3 \\ \hline \end{array}$$

c.
$$\begin{array}{r} 587 \\ \times\ 5.4 \\ \hline \end{array}$$

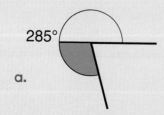

Part 8
Figure out each shaded angle.

a. 285°

b. 110°

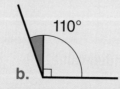

Lesson 71

Part 9 Work each problem.

a. There were 5 closed doors for every 7 doors in the building. There were 280 closed doors. How many doors were in the building? How many doors were open?

Part 9	
a.	

b. The ratio of gold coins to silver coins in a coin collection was 5 to 3. There were 240 coins in the collection. How many coins were silver? How many coins were gold?

Part 10 Copy each item. Show the repeated multiplication. Figure out the number it equals.

a. 2^4 b. 3^4 c. 1^4

Part 10	
a.	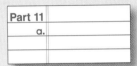

Part 11 Write each rounded value.

a. Round .4159 to the thousandths place.

b. Round .106 to the hundredths place.

c. Round .002 to the hundredths place.

d. Round .644 to the tenths place.

Part 11	
a.	

Lesson 72

Part 1
Copy and work each problem.

a. $7\overline{)1065}$
b. $3\overline{)863}$
c. $2\overline{)77}$
d. $5\overline{)728}$

Part 2
Copy each problem and complete the equation.

a. $135 \times 10^3 = $ ▬

d. $90 \times 10^3 = $ ▬

b. $70{,}200 = $ ▬ $\times$ ▬

e. $1{,}080{,}000 = $ ▬ $\times$ ▬

c. $28 \times 10^5 = $ ▬

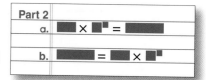

Part 3
Copy each number. Below, write the rounded number.

a. Round .2696 to the thousandths place.

b. Round .4093 to the thousandths place.

c. Round .5982 to the hundredths place.

d. Round .1297 to the thousandths place.

e. Round .946 to the tenths place.

Part 4
Work each problem.

a. $\frac{4}{10}$ of the windows were dirty. There were 60 windows. How many dirty windows were there?

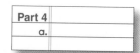

b. There were 2 dirty windows for every 7 windows. There were 32 dirty windows. How many windows were there?

c. $\frac{3}{5}$ of the ducks were swimming. There were 45 ducks in all. How many were swimming?

d. There were 2 open books for every 3 books. There were 16 open books. How many books were there in all?

Lesson 72

> ## Independent Work

Part 5 | Write an equation for each item.

    a. 10,000   **b.** 10^6 **c.** 100,000

Part 6 | Work each problem.

    a. $\frac{3}{4}$ of the students were reading. There were 48 students reading. How many students were there in all?

    b. There were 120 plants. $\frac{5}{6}$ of the plants had seeds. How many plants had seeds?

Part 7 | Find the volume of each retangular prism.

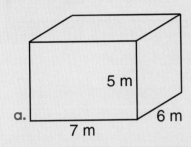 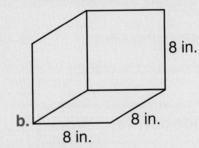

    a. 7 m, 6 m, 5 m

    b. 8 in., 8 in., 8 in.

Part 8 | Copy each problem and write the answer.

  a. 24 ÷ 4   **b.** 50 ÷ 5   **c.** 72 ÷ 9   **d.** 35 ÷ 7

  e. 42 ÷ 7   **f.** 81 ÷ 9   **g.** 30 ÷ 6   **h.** 28 ÷ 4

Part 9 | Write each rounded number.

    a. Round 38,824 to the thousands place.

    b. Round 259,302 to the ten-thousands place.

    c. Round 293,520 to the ten-thousands place.

Lesson 73

Part 1 Find the area of each circle.

$$\boxed{\begin{array}{c}\text{Area} = \text{pi}(\text{radius}^2) \\ A = \text{pi}(r^2)\end{array}}$$

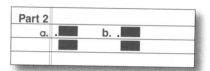

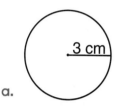

a.

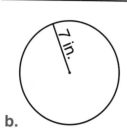

b.

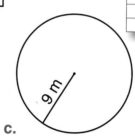

c.

Part 2 Copy each number. Below, write the rounded number.

a. Round .391 to the hundredths place.

b. Round .9862 to the tenths place.

c. Round .4959 to the hundredths place.

d. Round .72988 to the thousandths place.

Part 2			
a.	■	b.	■
	■		■

Independent Work

Part 3 Work each problem.

a. There were 5 red bugs for every 7 green bugs. There were 28 green bugs. How many red bugs were there?

Part 3	
a.	

b. $\frac{5}{6}$ of the ants were workers. There were 60 worker ants. How many ants were there in all?

c. Rachel walks 3 miles every 8 days. How many days will it take her to walk 15 miles?

d. $\frac{4}{9}$ of the pictures had frames. There were 72 pictures. How many pictures were framed?

Connecting Math Concepts

Lesson 73

Part 4
Copy and work each problem.

a.
$$\begin{array}{r} 5\,3\,7 \\ \times\ \ .4\,4 \\ \hline \end{array}$$

b.
$$\begin{array}{r} 8.7 \\ \times\ 9\,5 \\ \hline \end{array}$$

c.
$$\begin{array}{r} .4\,5\,6 \\ \times\ \ \ \ .6 \\ \hline \end{array}$$

Part 5
Copy and work each problem.

a. $3\overline{)7\,8\,3}$ b. $5\overline{)4\,2\,6}$ c. $4\overline{)5\,8\,4}$

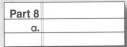

Part 6
Find the volume of each rectangular prism.

a.
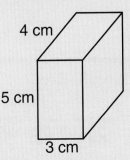
4 cm, 5 cm, 3 cm

b.
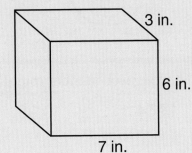
3 in., 6 in., 7 in.

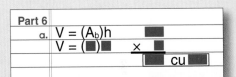

Part 7
Copy each item. Write the repeated multiplication. Figure out what it equals.

a. 5^3 b. 2^6 c. 3^4

Part 8
Find the area and perimeter of each figure.

a.
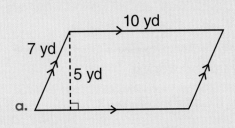
10 yd, 7 yd, 5 yd

b.
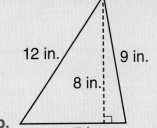
12 in., 9 in., 8 in., 7 in.

Part 9
Use estimation to work each problem.

a. $8\,3\overline{)2\,7\,8}$ b. $6\,9\overline{)5\,6\,0}$

Lesson 74

Part 1 — Find the area of each circle.

a.

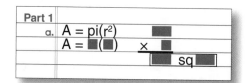

b.

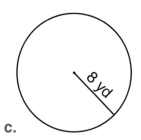

c.

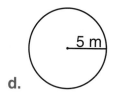

d.

Part 1				
a.	$A = pi(r^2)$		▦	
	$A = \blacksquare(\blacksquare)$	×	▦	
		▦	sq ▦	

Part 2 — Copy each number. Below, write each number rounded to the tenths place.

a. 4.98   b. 21.97   c. 1.954   d. 10.96

Part 2			
a.	▦	b.	▦
	▦		▦

Independent Work

Part 3 — Write the fraction that answers each question.

a. How many 12ths is $\frac{1}{3}$? $\frac{\blacksquare}{12}$ b. How many 10ths is $\frac{1}{5}$? $\frac{\blacksquare}{10}$

c. How many 14ths is $\frac{1}{2}$? $\frac{\blacksquare}{14}$ d. How many 20ths is $\frac{1}{4}$? $\frac{\blacksquare}{20}$

Part 3					
a.	▦	b.	▦	c.	▦
	▦		▦		▦

Part 4 — Write each problem with the decimal points lined up and work it.

a. 35.04 − 6.66   b. 70.35 + 24.4   c. 40 − 1.72

Part 4		
a.	▦.▦	
	▦.▦	
	.	

Lesson 74

Part 5 Find the volume of each rectangular prism.

a.

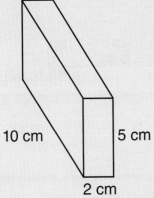

10 cm 5 cm 2 cm

b. 4 in.

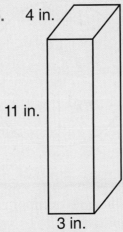

11 in. 3 in.

Part 5			
a.	$V = (A_b)h$	■	
	$V = (■)■$	× ■	
		■ cu ■	

Part 6 Work each problem.

a. A painter had 48 cans of paint. $\frac{3}{8}$ of the cans were full. How many cans were full?

b. There were 6 old trees for every 7 young trees. There were 126 young trees. How many old trees were there?

c. In a zoo, the ratio of red snakes to all snakes was 2 to 9. 56 snakes were not red. How many snakes were there in all? How many red snakes were there?

Part 6	
a.	

Part 7 Figure out the prime factors for each number.

a. 88   b. 66   c. 120   d. 126

Part 7		
a.	■	■

Part 8 Copy each letter equation. Replace one letter with a number. Figure out the other letter.

a. $\frac{4}{5}T = P$ ☐ T = 20

b. $\frac{8}{5}R = M$ ☐ M = 120

Part 8	
a.	■ ■ = ■
	■

Lesson 75

Part 1
Find the area of each circle.

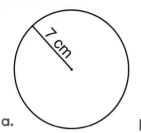 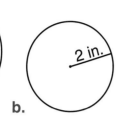 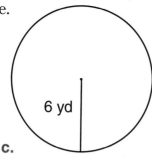

a. **b.** **c.**

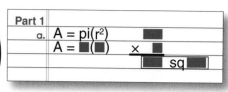

Part 1			
a.	$A = pi(r^2)$	■	
	$A = ■(■)$	×	■
		■ sq ■	

Part 2
Copy each number. Below, write each number rounded to the tenths place.

a. 42.95   **b.** 0.93   **c.** 163.96   **d.** 44.94   **e.** 8.983

Part 2		
a. ■		b. ■
■		■

Independent Work

Part 3
Answer each question.

a. How many degrees are in a circle?

b. How many degrees are in the corner of your paper?

c. How many degrees are in a straight line?

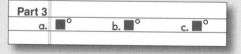

Part 3		
a. ■°	b. ■°	c. ■°

Part 4
Copy each item. Show the repeated multiplication. Figure out the number it equals.

a. 14^2 **b.** 3^3 **c.** 4^3 **d.** 10^4

Part 4	
a.	■■ = ■■■ = ■

Part 5
Use estimation to work each problem.

a. $54\overline{)296}$ **b.** $38\overline{)252}$

Part 5	
a.	■ ■■ ■■

Lesson 75

Part 6
Write the fraction that answers each question.

a. How many 8ths is $\frac{1}{2}$?

b. How many 30ths is $\frac{1}{6}$?

c. How many 12ths is $\frac{1}{4}$?

Part 7
Find the area and perimeter of each figure.

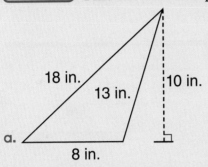

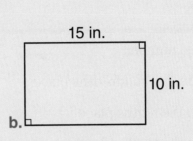

a. 18 in., 13 in., 10 in., 8 in.

b. 15 in., 10 in.

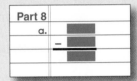

Part 8
Write each item as a column problem and work it.

a. 50 − 12.50

b. 70.10 − 68.81

c. 5009 − 349

Part 9
Work each problem.

a. At a meeting, the ratio of men to women was 3 to 2. There were 150 women at the meeting. How many men were at the meeting?

b. $\frac{2}{7}$ of the workers were resting. There were 22 workers resting. How many workers were there in all?

c. $\frac{3}{4}$ of the apples were ripe. There were 72 apples. How many apples were ripe?

d. The ratio of boys to girls at a picnic was 5 to 3. There were 80 children at the picnic. How many were boys? How many were girls?

Part 10
Copy and work each problem.

a. $3\overline{)586}$

b. $4\overline{)393}$

c. $7\overline{)447}$

Lesson 76

Part 1 Copy and work each problem.

a. $13\frac{7}{10}$
    $-11\frac{2}{10}$

b. $6\frac{1}{12}$
    $+4\frac{10}{12}$

c. $21\frac{3}{9}$
    $+ 8\frac{4}{9}$

d. $9\frac{5}{6}$
    $- 7\frac{3}{6}$

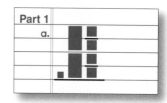

Part 2 Write each rounded number.

a. Round 30.698 to the hundredths place.

b. Round 4.925 to the tenths place.

c. Round 15.32 to the tenths place.

d. Round 1.099 to the hundredths place.

e. Round 145.087 to the tenths place.

f. Round 0.65 to the tenths place.

g. Round 406.958 to the hundredths place.

h. Round 73.925 to the tenths place.

i. Round 0.36951 to the thousandths place.

Part 2	
a.	▮

Part 3 Write the equation to show the mixed number.

a. $\frac{15}{11}$

b. $\frac{37}{30}$

c. $\frac{13}{7}$

d. $\frac{8}{5}$

Part 3	
a.	$\frac{15}{11} = 1\frac{\blacksquare}{\blacksquare}$

Lesson 76

Independent Work

Part 4
Work each problem.

a. In a swimming race $\frac{5}{8}$ of the swimmers are women. There are 120 women swimmers. How many swimmers are there in all? How many swimmers are men?

Part 4	
a.	

b. The ratio of old dogs to young dogs was 2 to 3. There were 150 dogs. How many were young? How many were old?

c. The ratio of hens to roosters was 3 to 5. There were 75 roosters. How many hens were there?

d. $\frac{3}{7}$ of the windows were clean. There were 42 windows. How many clean windows were there?

Part 5
Copy each item. Write the repeated multiplication. Figure out what it equals.

a. 12^2 b. 5^4 c. 9^3

Part 5	
a.	■■ = ■■■ = ■

Part 6
Copy each letter equation. Replace one letter with a number. Figure out the other letter.

a. $\frac{2}{7}T = V$ $\boxed{V = 56}$ b. $\frac{3}{4}B = C$ $\boxed{B = 48}$

Part 6	
a.	■ ■ = ■
	■

Part 7
Find the area of each circle.

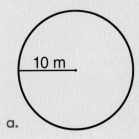

a.

Part 7	
a.	$A = pi(r^2)$
	$A = ■(■)$

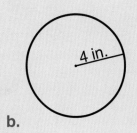

b.

Part 8 Write each rounded value.

    a. Round .8160 to the hundredths place.

    b. Round .5483 to the thousandths place.

    c. Round .3711 to the tenths place.

Part 8		
a.	b.	c.

Part 9 Write the fraction that answers each question.

a. How many 18ths is $\frac{1}{3}$?   **b.** How many 8ths is $\frac{1}{2}$?

c. How many 20ths is $\frac{1}{5}$?

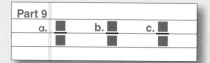

Part 9		
a. ▪	b. ▪	c. ▪
▪	▪	▪

Lesson 77

Part 1 For each division problem, write a multiplication problem and work it.

a. $\dfrac{2}{7} \div \dfrac{10}{14}$ b. $\dfrac{9}{10} \div \dfrac{6}{5}$ c. $\dfrac{20}{3} \div 4$ d. $\dfrac{1}{3} \div \dfrac{9}{4}$

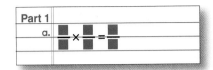

Part 2 Copy and work each problem.

a. $8\dfrac{9}{11}$ b. $15\dfrac{11}{18}$ c. $42\dfrac{7}{10}$ d. $8\dfrac{1}{9}$

$-\ 1\dfrac{7}{11}$ $+\ 3\dfrac{2}{18}$ $-\ 19\dfrac{3}{10}$ $+\ 25\dfrac{6}{9}$

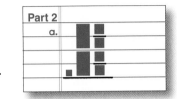

Part 3 Write the equation to show the mixed number.

a. $\dfrac{19}{10}$ b. $\dfrac{12}{7}$ c. $\dfrac{17}{15}$ d. $\dfrac{13}{9}$ e. $\dfrac{24}{23}$

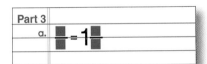

Independent Work

Part 4 Work each problem.

a. There were red flowers and white flowers in the garden. There were 4 red flowers for every 5 flowers. There were 64 red flowers. How many flowers were there in all? How many white flowers were there?

b. The ratio of brown turtles to green turtles was 9 to 7. There were 189 brown turtles. How many turtles were there? How many turtles were green?

Part 5 Copy and work each problem.

a. $4\overline{)388}$ b. $9\overline{)678}$ c. $5\overline{)743}$

Lesson 77

Part 6
Figure out the shaded angle.

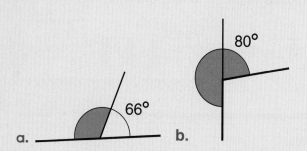

a. 66°

b. 80°

c. 15°

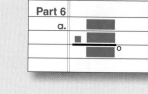

Part 6	
a.	▮▮
	▪ ▬
	▬°

Part 7
Figure out the prime factors for each number.

a. 99   b. 195   c. 145   d. 206

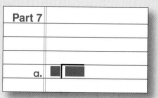

Part 7	
a.	▮ ▬

Part 8
Find the area of each circle.

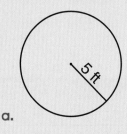 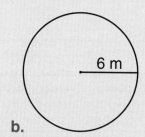

a. 5 ft

b. 6 m

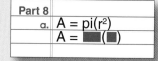

Part 8	
a.	A = pi(r²)
	A = ▮(▮)

Part 9
Use estimation to work each problem.

a. $33\overline{)269}$ b. $27\overline{)41}$

Part 9	
a.	▮ ▮▬ ▮ ▮

Part 10
Find the volume of each rectangular prism.

a.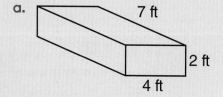

7 ft
2 ft
4 ft

b.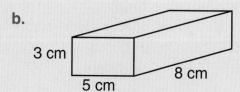

3 cm
5 cm
8 cm

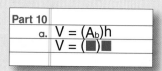

Part 10	
a.	V = (Aᵦ)h
	V = (▮)▮

Lesson 78

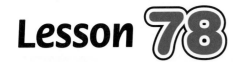

Work each problem. Answer the questions.

a. $\frac{2}{7}$ of the dogs were sleeping. 8 dogs were sleeping. How many dogs were awake? How many dogs were there?

b. There are 45 students. $\frac{4}{9}$ of the students are girls. How many girls are there? How many boys are there?

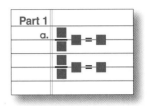

Part 2 For each division problem, write a multiplication problem and work it.

a. $\frac{1}{4} \div \frac{7}{8}$ b. $\frac{8}{5} \div \frac{1}{10}$ c. $\frac{9}{2} \div \frac{7}{3}$ d. $\frac{8}{9} \div 4$

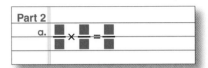

Part 3 Find the volume of each cylinder.

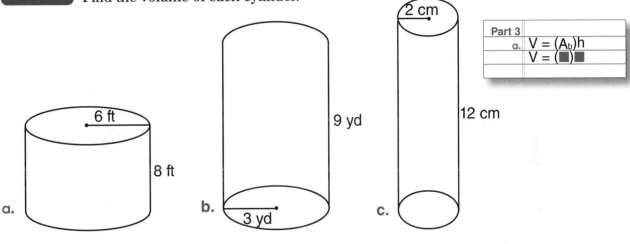

a. (cylinder: 6 ft radius, 8 ft height)

b. (cylinder: 3 yd radius, 9 yd height)

c. (cylinder: 2 cm radius, 12 cm height)

Independent Work

Part 4 Copy and work each problem.

a. $5\frac{3}{8}$
$-4\frac{2}{8}$

b. $12\frac{6}{10}$
$+31\frac{3}{10}$

c. $7\frac{3}{8}$
$-5\frac{3}{8}$

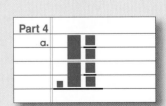

Lesson 78

Part 5
Find the area and perimeter of each figure.

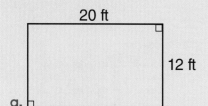

20 ft

12 ft

a.

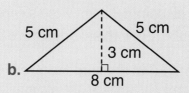

5 cm   5 cm

3 cm

b.

8 cm

Part 5	
a.	

Part 6
Work each problem.

a. There were 420 players in the league. 3 of every 10 players were experienced. How many players were not experienced? How many were experienced?

Part 6	
a.	

b. The ratio of tall men to short men at a meeting was 7 to 6. There were 210 tall men at the meeting. How many total men were at the meeting? How many short men were there?

Part 7
Write each item as a column problem and work it.

a. 600 − 354   b. 3.006 + 5.08   c. 2010 − 1006

Part 8
Copy each item. Write the repeated multiplication. Figure out what it equals.

a. 7^3 b. 3^5 c. 8^3

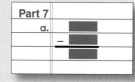

Part 9
Work each problem.

a. A board was $\frac{3}{4}$ feet long. Tom wanted to cut it so it was $\frac{1}{3}$ feet long. How many feet would he have to cut from the board?

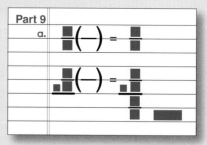

b. In a day, the dog ate $\frac{4}{5}$ pounds of the food and the cat ate $\frac{2}{9}$ pounds of food. How much total food did the two animals eat?

Lesson 79

Part 1 Work each problem.

a. There were 100 doors. $\frac{2}{5}$ of the doors were closed. How many open doors were there? How many closed doors were there?

b. $\frac{1}{6}$ of the birds were robins. There were 12 robins. How many birds were there? How many birds were not robins?

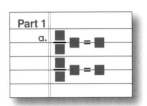

Part 2 Find the volume of each cylinder.

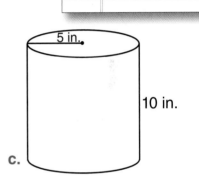

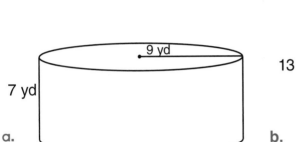

a. **b.** **c.**

Part 3 Copy each fraction. Write the simplified value it equals.

a. $\frac{38}{6}$ **b.** $\frac{50}{12}$ **c.** $\frac{4}{20}$ **d.** $\frac{56}{16}$

Independent Work

Part 4 Work each problem. $\boxed{S}$

a. $\frac{7}{5} \div \frac{2}{3}$ **b.** $\frac{9}{10} \div \frac{1}{5}$ **c.** $12 \div \frac{4}{9}$

Part 5 Write each rounded number.

a. Round .564 to the tenths place.

b. Round 10.468 to the hundredths place.

c. Round .0896 to the thousandths place.

Part 6 Copy and work each problem.

a. $14\frac{2}{9}$
$+ \ 6\frac{6}{9}$

b. $11\frac{5}{8}$
$- \ 6\frac{3}{8}$

c. $8\frac{12}{15}$
$- \ 7\frac{10}{15}$

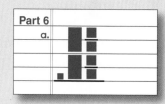

Part 7 Work each problem.

a. $\frac{3}{5}$ of the books were on the shelf. There were 30 books in all. How many books were on the shelf?

b. The ratio of paperback books to hardback books was 3 to 5. There were 56 books in all. How many were hardback books? How many were paperback books?

c. 6 out of every 8 books in the store were used. There were 20 new books. How many used books were there? How many books were there in all?

d. $\frac{1}{8}$ of the cards are red. There are 16 red cards. How many cards are there?

Part 8 Copy each fraction. Show the mixed number it equals.

a. $\frac{13}{12}$

b. $\frac{17}{15}$

c. $\frac{14}{9}$

d. $\frac{19}{10}$

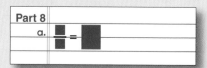

Part 9 Find the area of the circle. Start with the letter equation.

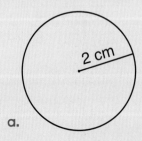

2 cm

a.

Lesson 80

Part 1 — Work each problem.

a. $\frac{2}{5}$ of the pies were baked. There were 14 baked pies. How many pies were not baked? How many pies were there in all?

b. The pie shop baked apple pies and blackberry pies. There were 60 pies in all. $\frac{3}{4}$ of the pies were apple pies. How many were blackberry pies? How many were apple pies?

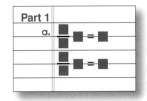

Part 1	
a.	■/■ ■ = ■
	■/■ ■ = ■

Part 2 — Find the volume of each figure.

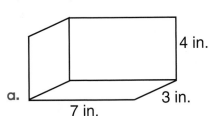

a. 7 in.   3 in.   4 in.

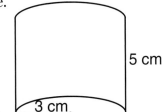

b. 3 cm   5 cm

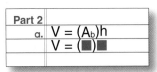

Part 2	
a.	$V = (A_b)h$
	$V = (■)■$

Part 3 — Copy and work each problem. Write the new addition below.

a. $13 + \frac{15}{11}$ b. $9 + \frac{23}{20}$ c. $7 + \frac{19}{10}$ d. $22 + \frac{18}{17}$

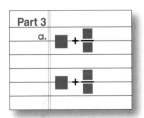

Part 3	
a.	■ + ■/■
	■ + ■/■

Part 4 — Copy each fraction. Write the simplified value it equals.

a. $\frac{18}{54}$ b. $\frac{57}{9}$ c. $\frac{37}{10}$ d. $\frac{75}{100}$

Part 4	
a.	■/■ =

Independent Work

Part 5 — Use estimation to work each problem.

a. $51\overline{)426}$ b. $76\overline{)464}$

Part 5	
a.	■ ■ ■ ■

Lesson 80

Part 6
Copy and work each problem. $\boxed{S}$

a. $24\frac{3}{5}$
$-\ 8\frac{3}{5}$

b. $2\frac{8}{15}$
$+\ 7\frac{6}{15}$

c. $21\frac{13}{14}$
$-\ 10\frac{11}{14}$

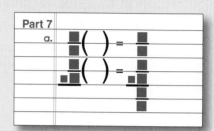

Part 7
Work each problem. $\boxed{S}$

a. Fran walked $\frac{3}{8}$ of a mile. Her sister walked $\frac{2}{3}$ of a mile. How far did the two girls walk in all?

b. Together, two kittens weighed $\frac{15}{4}$ pounds. One kitten weighed 2 pounds. How much did the other kitten weigh?

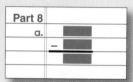

Part 8
Write each item as a column problem and work it.

a. 3640 − 1634

b. 80 − 2.46

c. 3007 − 998

Part 9
Copy each fraction. Show the mixed number it equals. $\boxed{S}$

a. $\frac{36}{25}$

b. $\frac{30}{21}$

c. $\frac{28}{18}$

d. $\frac{15}{14}$

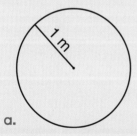

Part 10
Find the area of each circle.

a. (circle with radius 1 m)

b. (circle with 8 in.)

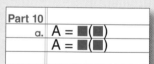

Part 11
Work each problem. $\boxed{S}$

a. $2 \div \frac{3}{2}$

b. $\frac{4}{9} \div \frac{5}{9}$

c. $\frac{13}{15} \div \frac{1}{2}$

Part 11	
a.	

Lesson 81

Part 1 Work each problem. Write the answer as a mixed number with a unit name. Ⓢ

a. A builder had a board that was $8\frac{3}{4}$ feet long. He cut it into two boards. One was $2\frac{1}{4}$ feet long. How long was the other board?

b. Ginger ran $7\frac{1}{8}$ miles on Monday and $3\frac{5}{8}$ miles on Tuesday. How far did she run on both days?

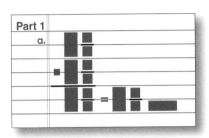

Part 2 Answer each question.

Temperatures at Noon in 3 Parks on Different Days	Monday	Tuesday	Wednesday
City Park	74°	76°	71°
Mountain Park	62°	59°	60°
River Park	71°	73°	71°

Part 2	
a.	

a. What was the lowest temperature on Monday?

b. Which park had the highest temperature on Tuesday?

c. How much colder was Wednesday than Tuesday at City Park?

d. What was the coldest temperature of the three-day period?

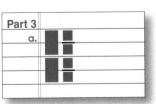

Part 3 Copy and rewrite each mixed number with a fraction less than 1.

a. $15\frac{11}{8}$ b. $4\frac{19}{15}$ c. $10\frac{14}{9}$ d. $26\frac{13}{12}$

Part 3	
a.	

Lesson 81

Independent Work

Part 4
Work the problem.

a. There were 90 students. $\frac{4}{5}$ of the students passed the test. How many students passed the test? How many failed the test?

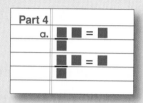

Part 5
Write each rounded number.

a. Round .568 to the hundredths place.

b. Round .853 to the tenths place.

c. Round .843 to the tenths place.

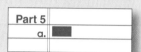

Part 6
Find the volume of each cylinder. Start with the letter equation.

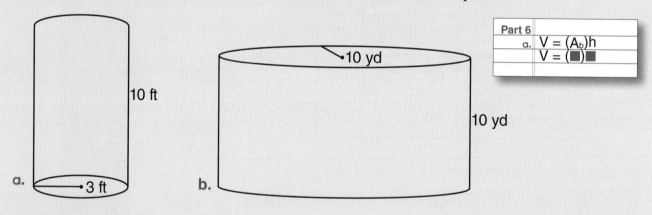

a. 10 ft, 3 ft

b. 10 yd, 10 yd

Part 6
a. $V = (A_b)h$
$V = (\blacksquare)\blacksquare$

Part 7
Copy each item. Write the repeated multiplication. Figure out what it equals.

a. 4^4 b. 5^3 c. 2^5

Part 7
a. $\blacksquare^\blacksquare = \blacksquare = \blacksquare$

Part 8
Work each problem. $\boxed{S}$

a. $\frac{3}{4} \div \frac{3}{4}$ b. $\frac{4}{7} \div 2$ c. $\frac{1}{8} \div \frac{5}{4}$ d. $\frac{5}{9} \div \frac{8}{3}$

Connecting Math Concepts

Lesson

Part 1 Work each problem. Write the answer as a mixed number with a unit name.

a. Tom picked $13\frac{2}{5}$ pounds of apples. Al picked $7\frac{2}{5}$ pounds of apples. How many pounds of apples did the men have together?

b. The dog weighed $18\frac{11}{16}$ pounds. The dog lost $2\frac{10}{16}$ pounds. How much did the dog weigh then?

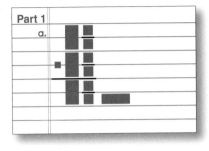

Part 2 Copy and rewrite each mixed number with a fraction that is less than 1.

a. $9\frac{27}{20}$ b. $15\frac{17}{12}$ c. $2\frac{13}{10}$ d. $1\frac{19}{13}$ e. $19\frac{7}{5}$

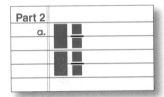

Part 3 Write the fraction that answers each question.

a. $\frac{4}{5}$ of the glasses were dirty.

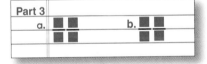

- What's the fraction for all the glasses?

- What's the fraction for clean glasses?

b. $\frac{11}{15}$ of the bricks were stacked.

- What fraction of the bricks were not stacked?

- What's the fraction for all the bricks?

c. $\frac{3}{8}$ of the horses were for sale.

- What's the fraction for all the horses?

- What's the fraction for horses that were not for sale?

Lesson 82

Part 4 Find the area of each circle.

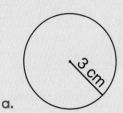

a.

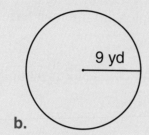

9 yd

b.

Part 4		
a.	A = ■(■²)	■
	A = ■(■²)	× ■
		■

Part 5 Copy and work each problem.

a. 6⟌379   b. 8⟌586   c. 6⟌433

Part 5	
a.	■ ■

Part 6 Copy each fraction. Show the mixed number it equals.

a. $\dfrac{23}{12}$ b. $\dfrac{31}{21}$ c. $\dfrac{11}{9}$ d. $\dfrac{46}{43}$

Part 6	
a.	■ = ■ ■

Part 7 Find the volume of each figure. Start with the letter equation.

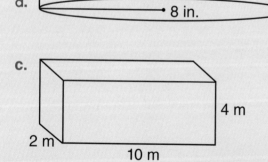

a.   • 8 in.

2 in.

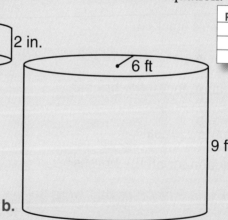

6 ft

9 ft

b.

Part 7	
a.	V = (A_b)h
	V = (■)■

c.

4 m

2 m   10 m

Part 8 Work each problem. ⑤

a. $\dfrac{2}{10} \div \dfrac{1}{8}$ b. $3 \div \dfrac{11}{5}$ c. $\dfrac{17}{5} \div \dfrac{3}{10}$ d. $\dfrac{4}{9} \div \dfrac{4}{9}$

Part 8	
a.	

Lesson 82

Part 9 Work each problem.

a. $\frac{2}{3}$ of the windows were open. There were 14 open windows. How many windows were there in all? How many windows were closed?

b. $\frac{1}{4}$ of the animals were sleeping. There were 20 animals. How many animals were awake? How many animals were asleep?

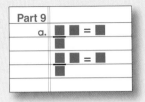

Lesson

Part 1 Find the average.

Number of Pounds Different Boxes Weigh

a.

22	45	12	50

Temperatures on Different Days

b.

Monday	Tuesday	Wednesday	Thursday	Friday
38°	27°	22°	33°	40°

Part 1	
a.	

Part 2 Write the fraction that answers each question.

a. $\frac{1}{4}$ of the eggs were brown.

- What's the fraction for all the eggs?

- What's the fraction for eggs that were not brown?

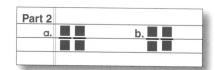

Part 2		
a.	▪▪ ▪▪	b. ▪▪ ▪▪

b. $\frac{11}{12}$ of the windows were open.

- What fraction of the windows were closed?

- What's the fraction for all the windows?

c. $\frac{40}{100}$ of the eggs were large.

- What fraction of the eggs were small?

- What's the fraction for all the eggs?

Connecting Math Concepts

Lesson 83

Part 3 Work each problem. Write the answer as a mixed number with a unit name. S

a. Mrs. Green started out with $7\frac{3}{8}$ pounds of salad. She served $5\frac{1}{8}$ pounds of salad. How many pounds of salad did she still have?

b. Last year, the state paved $14\frac{5}{8}$ miles of Spencer Road. This year, the state paved $43\frac{2}{8}$ miles of Spencer Road. How many miles of road did the state pave in both years?

c. The tan dog could jump $26\frac{3}{5}$ inches. The brown dog could jump $28\frac{4}{5}$ inches. How much higher could the brown dog jump than the tan dog?

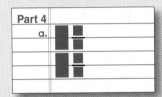

Part 4 Copy and rewrite each mixed number. S

a. $3\frac{9}{8}$ b. $12\frac{14}{8}$ c. $6\frac{5}{5}$

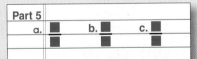

Part 5 Write the fraction that answers each question.

a. How many 15ths is $\frac{1}{3}$?   d. How many 16ths is $\frac{1}{4}$?

b. How many 12ths is $\frac{1}{2}$?   e. How many 16ths is $\frac{1}{2}$?

c. How many 10ths is $\frac{1}{5}$?

Part 6 Work each problem.

a. $\frac{4}{9} \div \frac{9}{4}$ b. $7 \div \frac{4}{5}$ c. $\frac{15}{6} \div 2$

Lesson

Part 7 Work each problem.

a. $\frac{3}{4}$ of the flowers were red. There were 27 red flowers. How many flowers were not red? How many flowers were there in all?

Part 7	
a.	

b. $\frac{4}{7}$ of the windows were open. There were 49 windows. How many were closed? How many were open?

c. There were cows and pigs on the farm. The ratio of cows to all animals was 3 to 10. There were 140 pigs. How many cows were there? How many animals were there?

Connecting Math Concepts

Lesson 84

Part 1 Figure out what each item equals.

a. $8(12 \div 2)$

b. $10(22 + 3 - 5)$

c. $3\left(\dfrac{8}{9} - \dfrac{7}{9}\right)$

d. $12(5 \times 2)$

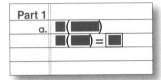

Part 2 Work each problem. $\boxed{S}$

a. A doctor divides $\dfrac{7}{8}$ ounce of medicine into 5 equal doses. How many ounces is each dose?

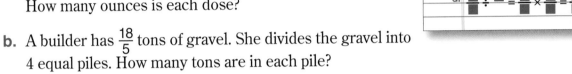

b. A builder has $\dfrac{18}{5}$ tons of gravel. She divides the gravel into 4 equal piles. How many tons are in each pile?

Part 3 Find the average height of each team.

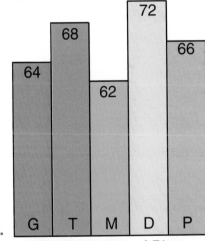

a.

Height in Inches of Players
on the Rockets Team

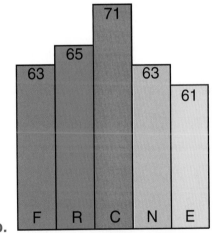

b.

Height in Inches of Players
on the Fliers Team

Lesson 84

Part 4 Work each problem. $\boxed{S}$

a. The new table is $\frac{9}{2}$ feet long. The old table is 6 feet long. How much longer is the old table than the new table?

b. Dan worked $\frac{11}{2}$ hours. Tom worked $\frac{8}{3}$ hours. How many hours did they work altogether?

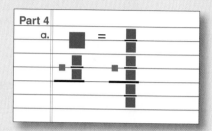

Part 5 Find the volume of each figure. Start with the letter equation.

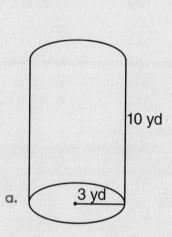

a.

10 yd

3 yd

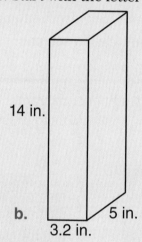

14 in.

b. 5 in.

3.2 in.

Part 5
a.

Part 6 Use estimation to work each problem.

a. $3\,1\,\overline{)1\,6\,5}$

b. $4\,1\,\overline{)3\,4\,1}$

Part 7 Write each item as a column problem and work it.

a. 42 − 9.51

b. 3.8 × .02

c. 10.4 − 9.03

d. 9.05 × 1.5

e. 3006 − 714

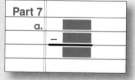

Lesson 84

Part 8 Work each problem.

a. $\frac{2}{5}$ of the boxes in the room were empty. There were 30 empty boxes. How many boxes were there in all? How many boxes were not empty?

Part 8	
a.	

b. The ratio of trees to bushes was 4 to 3. There were 24 trees. How many plants were there? How many bushes were there?

c. $\frac{3}{8}$ of the dogs were barking. There were 24 dogs. How many dogs were not barking? How many dogs were barking?

d. On the Dodge Building, there were 5 dirty windows for every 4 clean windows. There were 270 windows in all. How many dirty windows were there? How many clean windows were there?

Part 9 Copy and work each problem.

a. $8\overline{)928}$ b. $5\overline{)476}$ c. $8\overline{)184}$ d. $7\overline{)565}$

Part 9	
a.	

Lesson 85

Part 1   Write the answer to each division problem.

a. $\dfrac{119.08}{10^2}$ b. $\dfrac{382}{10^3}$ c. $\dfrac{12.06}{10^3}$ d. $\dfrac{.5}{10^2}$ e. $\dfrac{10.03}{10^1}$

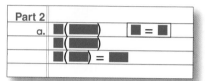

Part 2   Copy each item. Replace the letter with a number and work the problem.

a. $5(4 + 2 - r)$ $\boxed{r = 1}$ b. $2\left(\dfrac{3}{12} + n\right)$ $\boxed{n = \dfrac{4}{12}}$

c. $m(12 \div 4)$ $\boxed{m = 9}$ d. $5(13 - k + 12)$ $\boxed{k = 5}$

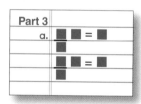

Part 3   Write two equations for each sentence.

a. $\dfrac{2}{5}$ of the cats were black.

b. $\dfrac{4}{7}$ of the plates were wet.

c. $\dfrac{1}{9}$ of the bolts were rusty.

Independent Work

Part 4   Work each problem. Write the answer as a mixed number with a unit name. $\boxed{S}$

a. Last year, Mr. Gregory weighed $178\frac{2}{16}$ pounds. During the year, he gained $12\frac{5}{16}$ pounds. How much does he weigh now?

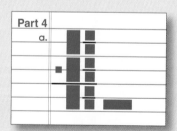

b. The turtle was $7\frac{5}{12}$ years old. The dog was $10\frac{11}{12}$ years old. How much older was the dog than the turtle?

c. Josh ran $16\frac{1}{5}$ miles over the weekend. Karen ran $18\frac{2}{5}$ miles. How many total miles did the two runners run?

Lesson 85

Part 5 Write each rounded value.

a. Round .3608 to the tenths place.

b. Round .0483 to the hundredths place.

c. Round .7654 to the thousandths place.

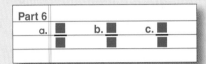

Part 5			
a.	b.	c.	

Part 6 Write the fraction that answers each question.

a. How many 18ths is $\frac{1}{6}$?   b. How many 24ths is $\frac{1}{3}$?

c. How many 12ths is $\frac{1}{4}$?

Part 7 Copy and work each problem.

a. $6\overline{)154}$ b. $3\overline{)599}$ c. $8\overline{)470}$ d. $9\overline{)511}$

Part 7	
a.	

Part 8 Work each problem.

a. $\frac{2}{3}$ of the people were hungry. There were 18 people. How many were hungry? How many were not hungry?

Part 8	
a.	

b. $\frac{1}{5}$ of the workers rode bikes, the rest drove cars. 35 workers rode bikes. How many workers were there? How many drove cars?

c. The ratio of wet fields to dry fields was 6 to 5. There were 121 fields. How many fields were wet? How many fields were dry?

Part 9 Work each problem. $\boxed{S}$

a. $\frac{16}{4} \div 4$ b. $\frac{3}{8} \div \frac{1}{10}$ c. $\frac{12}{7} \div 3$

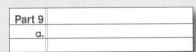

Part 9	
a.	

Part 10 Copy and rewrite each mixed number.

a. $18\frac{14}{10}$ b. $7\frac{8}{5}$ c. $3\frac{18}{15}$

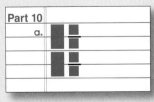

Lesson 86

Part 1 — Write the answer to each problem.

a. 4.6×10^3

b. $\dfrac{4.6}{10^3}$

c. $\dfrac{110.8}{10^2}$

d. 6.02×10^4

e. $\dfrac{285}{10^4}$

Part 2 — Work each problem. Write the answer as a number and a unit name.

a. A farmer has $\dfrac{8}{3}$ quarts of honey. He wants to put that honey in jars that each hold $\dfrac{2}{3}$ quart. How many jars will he be able to fill?

b. Another farmer has $\dfrac{9}{4}$ quarts of honey. He wants to fill jars that each hold $\dfrac{2}{3}$ quart. How many jars will he fill?

c. Another farmer has 8 quarts of honey. She wants to fill jars that each hold $\dfrac{2}{3}$ quart. How many jars will she fill?

Part 3 — Figure out what each item equals.

a. $t(15 \div 3)$ $\boxed{t = 10}$

b. $8\left(r + \dfrac{3}{5}\right)$ $\boxed{r = \dfrac{1}{5}}$

c. $k(12 + 5 - 7)$ $\boxed{k = \dfrac{1}{2}}$

d. $m(15 - 5 - 4)$ $\boxed{m = 20}$

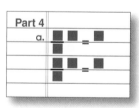

Part 4 — Write two equations for each sentence.

a. $\dfrac{2}{9}$ of the men were sleeping.

b. $\dfrac{15}{19}$ of the girls were eating.

c. $\dfrac{3}{16}$ of the computers were used.

d. $\dfrac{10}{11}$ of the papers were blank.

Lesson

Independent Work

Part 5 Copy each fraction. Show the mixed number it equals.

a. $\dfrac{31}{24}$ b. $\dfrac{15}{11}$ c. $\dfrac{47}{35}$ d. $\dfrac{30}{16}$

Part 6 Work each problem.

a. There were 2 short pencils for every 7 long pencils. There were 63 long pencils. How many pencils were there? How many pencils were short?

b. $\dfrac{2}{3}$ of the people at the beach were standing. There were 90 people at the beach. How many people were standing? How many people were not standing?

c. $\dfrac{3}{4}$ of the hotdogs had mustard on them. 36 hotdogs had mustard on them. How many hotdogs were there? How many did not have mustard?

Part 7 Find the volume of each figure. Start with the letter equation.

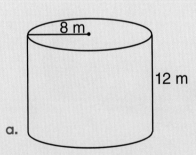

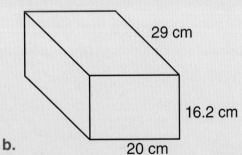

a.

b.

Part 8 Copy and work each problem.

a. $14\dfrac{4}{5}$ b. $7\dfrac{2}{9}$ c. $11\dfrac{2}{3}$
  $+\ 8\dfrac{3}{5}$ $+3\dfrac{7}{9}$ $+10\dfrac{2}{3}$

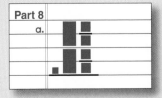

Lesson 86

Part 9 Work each problem.

a. Pat bought a mirror and a chair. The mirror cost $34.56. The chair cost $58.98. How much did both pieces of furniture cost?

Part 9	
a.	

b. Barb was $65\frac{3}{4}$ inches tall. Val was $64\frac{1}{4}$ inches tall. How much taller was Barb than Val?

c. The apple tree was $17\frac{11}{12}$ feet tall last year. The tree grew $4\frac{5}{12}$ feet. How tall is the tree now?

Part 10 Copy each item. Write the repeated multiplication. Figure out what it equals.

a. 5^5 b. 2^5 c. 12^2

Part 10	
a.	■■ = ■■ = ■

Part 11 Copy and work each problem.

a. $5\overline{)360}$ b. $8\overline{)360}$ c. $6\overline{)989}$ d. $5\overline{)745}$

Part 11	
a.	■ ■

Lesson

Part 1 Find the average for each month.

Number of Chin-ups		
	Last Month	This Month
Jim	4	4
Fran	0	1
Alex	0	0
Liz	8	7
Paul	18	13

Part 1	
a.	

Part 2 Work each problem. Write the answer as a number and a unit name.

a. A store owner has $\frac{9}{4}$ pounds of syrup. She pours the syrup into jars that each hold $\frac{3}{8}$ pound. How many jars can she fill?

Part 2	
a.	

b. Fran has 26 pounds of walnuts. She wants to put them in bags that hold $\frac{3}{4}$ pound. How many bags can she fill?

c. A tank held 35 gallons of gasoline. The gasoline was poured into containers that each held $\frac{2}{3}$ gallon. How many containers were filled?

Lesson 87

Part 3   Write two equations for each sentence.

a. $\frac{3}{4}$ of the fish were alive.

 • Write the equation for fish that were alive and the equation for fish that were dead.

b. $\frac{7}{8}$ of the boxes are upstairs.

 • Write the equation for the boxes upstairs and the equation for the boxes downstairs.

c. $\frac{2}{3}$ of the papers are white.

 • Write the equation for the papers that are white and the equation for the papers that are not white.

d. $\frac{5}{9}$ of the animals were hungry.

 • Write the equation for the animals that were hungry and the equation for the animals that were not hungry.

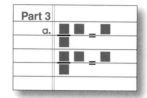

Independent Work

Part 4   Write each item as a column problem and work it.

a. 4006 − 1247   b. 4818 + 3919   c. 8.02 × .7

d. 3.99 × .05   e. 35 − 24.99

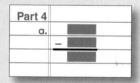

Part 5   Answer each question.

Number of People Visiting a Campground Each Day					
Sunday	Monday	Tuesday	Wednesday	Thursday	Friday
30	44	70	11	99	100

a. What is the average number of people that visited the campground each day?

b. How many days are above the average?

c. How many days are below the average?

Part 6 Write the fraction that answers each question.

a. How many 20ths is $\frac{1}{5}$? b. How many 12ths is $\frac{1}{4}$?

c. How many 24ths is $\frac{1}{2}$?

Part 6			
a. ■	b. ■	c. ■	
■	■	■	

Part 7 Write each rounded value.

a. Round .5438 to the thousandths place.

b. Round .1950 to the hundredths place.

c. Round 4.6615 to the tenths place.

Part 7		
a.	b.	c.

Part 8 Work each problem.

a. The ratio of dull knives to all knives was 2 to 5. There were 85 knives. How many sharp knives were there? How many dull knives were there?

b. $\frac{4}{9}$ of the birds at the beach were white. There were 18 birds. How many birds were not white? How many birds were white?

c. The ratio of all cats to purring cats is 9 to 1. There are 36 cats. How many are not purring? How many are purring?

Part 8	
a.	

Part 9 Figure out what each item equals.

a. $4(k \div 4)$ $\boxed{k = 48}$

b. $3(5 + 7 - B)$ $\boxed{B = 11}$

c. $t(10 - 7 + 3)$ $\boxed{t = 5}$

Part 10 Copy and work each problem.

a. $8\overline{)736}$ b. $5\overline{)255}$ c. $4\overline{)938}$ d. $7\overline{)655}$

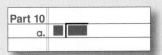

Lesson 87

Part 11 Work each problem. Write the answer as a mixed number with a unit name.

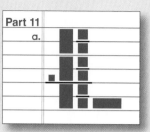

a. Dan slept for $10\frac{7}{8}$ hours. His mom slept for $6\frac{7}{8}$ hours. How much longer did Dan sleep than his mom?

b. Ted read $46\frac{1}{4}$ pages of the book yesterday. He read $20\frac{3}{4}$ pages today. How much did he read on both days?

c. Rob collected $6\frac{5}{8}$ pounds of apples. Jose collected $10\frac{6}{8}$ pounds of apples. How many pounds of apples did they collect in all?

Part 12 Write the answer to each problem.

a. $\dfrac{3.6}{10^1}$

b. $\dfrac{42.3}{10^3}$

c. $\dfrac{.03}{10^2}$

Part 12	
a.	

d. 1.8×10^2

e. $.005 \times 10^3$

f. 5.16×10^2

Connecting Math Concepts

Lesson

Part 1 Work each problem and write the answer.

a. Show the quantity 300 ÷ 6. Then multiply by 7.

b. Show the quantity 27 – 7 – 10. Then multiply by 8.

c. Show the quantity 65 – 40. Then multiply by 4.

d. Show the quantity 80 ÷ 2. Then multiply by $\frac{1}{2}$.

e. Show the quantity 7 + 3 – 5. Then multiply by 30.

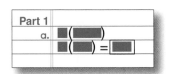

Part 2 Work each problem. Write the simplified answer and a unit name.

a. There are $\frac{28}{3}$ gallons of paint. The paint is divided equally into 4 containers. How many gallons are in each container?

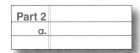

b. There are $\frac{28}{3}$ gallons of paint. The paint is poured into containers that each hold $\frac{4}{3}$ gallons. How many containers will be filled?

c. $\frac{7}{4}$ pounds of shrimp make 4 servings. How much does each serving weigh?

d. $\frac{8}{3}$ pounds of shrimp is divided into servings that each weigh $\frac{2}{5}$ pound. How many servings can be made?

Lesson 88

Part 3 Find the average distance.

Part 3	
a.	

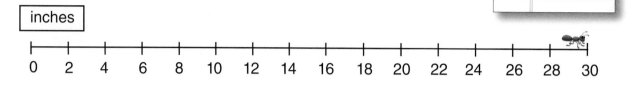

inches

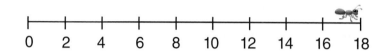

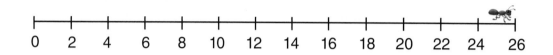

a.

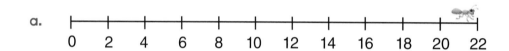

M ————————————————

P —————————————————————

R ——————————————————

T ——————————

b.

centimeters

Part 4 Write the letter equation. Below, write the equation with a number.

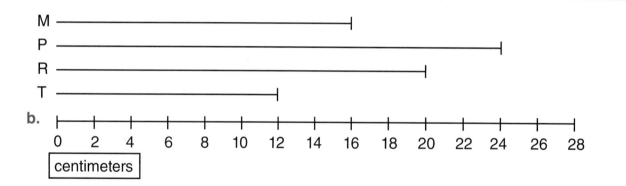

a. $\frac{3}{8}$ of the children were eating. 40 children were not eating.

b. $\frac{2}{3}$ of the books are old. There are 24 new books.

c. $\frac{4}{9}$ of the students were girls. There were 360 boys.

d. $\frac{2}{5}$ of the animals were not eating. 15 animals were eating.

e. $\frac{1}{10}$ of the eggs were broken. There were 27 unbroken eggs.

Lesson 88

Independent Work

Part 5 Figure out what each item equals.

a. $12(1 + K - 4)$ $\boxed{K = 6}$

b. $T(4 - 1 + 5)$ $\boxed{T = 2}$

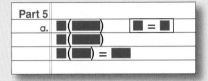

Part 6 Copy and work each problem.

a. $7\overline{)408}$ b. $3\overline{)306}$ c. $5\overline{)489}$ d. $6\overline{)789}$

Part 7 Work each problem.

a. The crew painted $\frac{2}{5}$ of the rooms of a hotel. 80 rooms were painted. How many rooms are in the hotel? How many rooms are not painted?

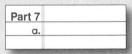

b. There was 1 open door for every 8 doors. 42 doors were closed. How many doors were there? How many doors were open?

c. $\frac{3}{8}$ of the children walked. The rest rode the bus. There were 48 children. How many rode the bus? How many walked?

Part 8 Copy and work each problem.

a. $2\frac{16}{17}$ b. $5\frac{7}{9}$ c. $18\frac{3}{5}$

$+ 4\frac{2}{17}$ $+ 6\frac{4}{9}$ $+ 12\frac{2}{5}$

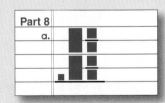

Lesson 88

Part 9 Write two equations for each item.

a. $\dfrac{2}{7}$ of the students were noisy.

b. $\dfrac{1}{12}$ of the pencils were broken.

c. $\dfrac{11}{20}$ of the children were sleeping.

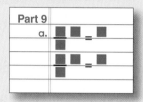

Part 10 Write the answer to each problem.

a. $\dfrac{.03}{10^2}$

b. $\dfrac{56}{10^3}$

c. $\dfrac{4.813}{10^1}$

d. 4.6×10^1

e. $.002 \times 10^2$

f. 5.13×10^3

Part 11 Copy and work each problem.

a. $8\,2\overline{)6\,5\,7}$

b. $4\,6\overline{)2\,4\,3}$

Lesson 89

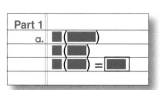

Part 1   Work each problem.

a. Add $\frac{3}{2}$ and $\frac{7}{2}$. Multiply by 9.

b. Add 80 and 10. Subtract 60. Then multiply the quantity by 2.

c. Add 15 and 11. Subtract 6. Then multiply by 5.

d. Subtract $\frac{4}{6}$ from $\frac{40}{6}$. Then multiply by 3.

e. Add 12 and 12. Subtract 10. Then multiply by 10.

Part 2   Work each problem. Write the answer as a number and a unit name.

a. Robert wants to put raisins into bags that each hold $\frac{7}{8}$ pound. He has 42 pounds of raisins. How many bags can be filled?

b. Each dose of medicine weighs $\frac{1}{18}$ ounce. How many doses can be made from $\frac{2}{3}$ ounce of medicine?

c. $\frac{35}{2}$ cups of flour make 7 cake recipes. How many cups of flour does each recipe call for?

d. A pie is divided into 9 slices. The pie weighs $\frac{81}{2}$ ounces. How many ounces does each slice weigh?

Lesson 89

Part 3 — Find the average.

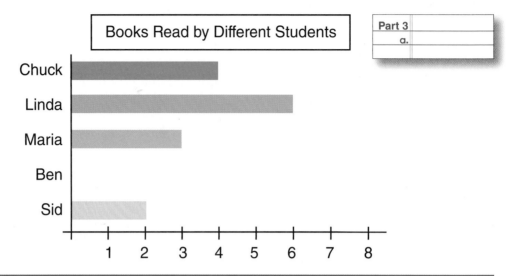

Books Read by Different Students

Part 3	
a.	

a.

Visits to a Doctor Last Year

0	4	0	12	4
Jamie	Billy	Kathy	Sally	Roger

b.

Part 4 — Write the letter equation. Below, write the equation with a number.

a. $\frac{3}{4}$ of the meals are cooked. There are 48 uncooked meals.

b. $\frac{7}{10}$ of the assignments are not corrected. 30 assignments are corrected.

c. $\frac{1}{9}$ of the horses are male. 72 of the horses are female.

d. $\frac{8}{15}$ of the apples are eaten. 56 apples are not eaten.

e. $\frac{13}{20}$ of the hats are clean. There are 7 dirty hats.

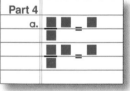

Part 4	
a.	

Independent Work

Part 5 — Work each problem.

a. $\frac{12}{5} \div 4$ b. $\frac{3}{12} \div \frac{1}{2}$ c. $5 \div \frac{6}{9}$

Part 5	
a.	

Part 6 Figure out what each item equals.

a. 5(B ÷ 6)   B = 60

b. R(3 + 9)   R = 3

Part 7 Find the volume of each figure. Start with the letter equation.

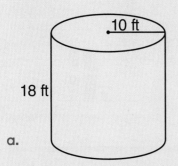

a.

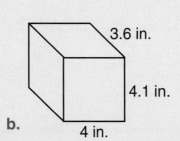

b.

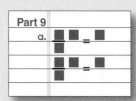

Part 8 Work each problem.

a. $\frac{3}{4}$ of the towels were old. There were 68 towels. How many were new? How many were old?

b. The ratio of moving cars to parked cars was 9 to 2. There were 80 parked cars. How many cars were there? How many moving cars were there?

c. There were 3 coins for every 2 bills. There were 90 items in all. How many bills were there? How many coins were there?

Part 9 Write two letter equations for each item.

a. $\frac{3}{8}$ of the shoes were dirty.

b. $\frac{12}{17}$ of the doors were open.

c. $\frac{2}{9}$ of the towels were wet.

Lesson 89

Part 10 Copy and complete each equation.

a. $\dfrac{13.1}{10^2} = $ ▮

b. $\dfrac{2.13}{10^2} = $ ▮

c. $\dfrac{48}{10^3} = $ ▮

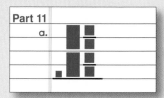

Part 10	
a.	

d. $.009 \times 10^2 = $ ▮

e. $3.14 \times 10^1 = $ ▮

f. $.461 \times 10^3 = $ ▮

Part 11 Copy and work each problem.

a. $13\dfrac{4}{9}$
$+ 20\dfrac{8}{9}$

b. $12\dfrac{9}{15}$
$+ 5\dfrac{8}{15}$

c. $1\dfrac{3}{10}$
$+ \dfrac{9}{10}$

Part 12 Copy and work each problem.

a. $8\overline{)278}$

b. $7\overline{)309}$

c. $9\overline{)309}$

d. $6\overline{)493}$

e. $7\overline{)512}$

f. $7\overline{)362}$

Lesson

Part 1 Copy and work each problem.

a.
$$9$$
$$- 2\frac{3}{4}$$

b.
$$7\frac{4}{8}$$
$$- 3\frac{5}{8}$$

c.
$$14\frac{2}{7}$$
$$- 2\frac{5}{7}$$

d.
$$10$$
$$- 3\frac{2}{5}$$

Part 1	
a.	

Part 2 Work each problem. Write the answer as a number and a unit name.

a. How many servings of $\frac{3}{4}$ pound each would somebody get from 6 pounds of salad?

b. Bill has $\frac{45}{7}$ gallons of cream. He wants to fill 5 containers. How much cream will be in each container?

c. You want to make 3 servings from $\frac{55}{2}$ ounces of salad. How many ounces would be in each serving?

d. Each container holds $\frac{7}{8}$ pound of nuts. How many containers can be filled with 14 pounds of nuts?

Part 2	
a.	

Independent Work

Part 3 Answer the question.

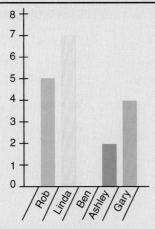

Hours Students Spent on Homework Last Week

a. What is the average number of hours spent on homework?

Part 3	
a.	

Lesson 90

Part 4 Figure out what each item equals.

     a. $10(P \div 8)$ $\boxed{P = 32}$

     b. $J(15 - 3)$ $\boxed{J = 6}$

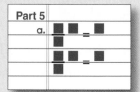

Part 5 Write two letter equations for each item.

     a. $\frac{1}{8}$ of the dogs were barking.

     b. $\frac{5}{13}$ of the plates were dry.

     c. $\frac{1}{4}$ of the students were reading.

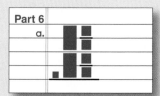

Part 6 Copy and work each problem.

a. $\begin{array}{r} 20\frac{7}{12} \\ +\,50\frac{5}{12} \\ \hline \end{array}$

b. $\begin{array}{r} 3\frac{15}{18} \\ +\,7\frac{5}{18} \\ \hline \end{array}$

c. $\begin{array}{r} \frac{7}{8} \\ +\,12\frac{3}{8} \\ \hline \end{array}$

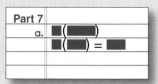

Part 7 Write each problem and work it.

     a. Subtract $\frac{3}{4}$ from $\frac{7}{4}$. Then multiply by 9.

     b. Add $\frac{3}{5}$ and $\frac{9}{5}$. Then multiply by 4.

     c. Add 18 and 54. Then multiply by $\frac{1}{4}$.

     d. Subtract $\frac{3}{8}$ from $\frac{9}{8}$. Then multiply by 8.

Lesson 90

Part 8 Work each problem.

a. There were men and women at the meeting. The ratio of men to women was 4 to 9. There were 52 men. How many people were there? How many women were there?

Part 8	
a.	

b. There were 210 cartons in the storage room. The ratio of opened cartons to all cartons was 3 to 7. How many cartons were opened? How many were unopened?

Part 9 Copy and work each problem.

a. $7\overline{)406}$ b. $6\overline{)345}$ c. $8\overline{)987}$

d. $7\overline{)221}$ e. $6\overline{)466}$ f. $7\overline{)690}$

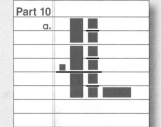

Part 9	
a.	

Part 10 Work each problem. Write the answer as a mixed number with a unit name.

a. Hank weighed $3\frac{1}{4}$ pounds less than Al. Al weighed $128\frac{3}{4}$ pounds. How much did Hank weigh?

b. In Lincoln City, it rained $2\frac{3}{8}$ inches on Monday, and $3\frac{1}{8}$ inches on Tuesday. How much did it rain for both days?

c. The board was $3\frac{2}{10}$ inches shorter than the table. The table was $69\frac{7}{10}$ inches long. How long was the board?

Part 10	
a.	

Lesson

Part 1   Work each problem. Remember the unit name.

a. 1 week is 7 days. Change 11 weeks into days.

b. 1 year is 12 months. Change 10 years into months.

c. 1 hour is 60 minutes. Change 8 hours into minutes.

Part 1	
a.	■ × ■ = ■■

Part 2   Work each problem. Write the answer as a number and a unit name.

a. Mary uses $\frac{18}{5}$ pounds of potatoes to make mashed potatoes. She needs to make 6 meals. How many pounds is each serving of mashed potatoes?

b. Each perfume bottle holds $\frac{2}{5}$ ounce. How many bottles can be filled if there are 4 ounces of perfume?

c. How many $\frac{5}{8}$-pound servings of shrimp can be made from 10 pounds?

d. A whole pizza weighs $\frac{7}{2}$ pounds. The pizza is cut into 8 equal slices. What does each slice weigh?

Part 2	
a.	

Part 3   Work each problem.

a. $\frac{3}{4}$ of the flowers were roses. 28 flowers were not roses. How many flowers were there? How many roses were there?

b. $\frac{3}{10}$ of the doors were open. There were 28 closed doors. How many doors were there? How many doors were open?

c. $\frac{4}{7}$ of the plates were dry. There were 21 wet plates. How many dry plates were there? How many plates were there in all?

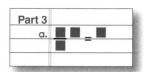

Independent Work

Part 4   Copy and work each problem.

a. $5\frac{2}{9}$
$-2\frac{7}{9}$

b. 7
$-3\frac{2}{5}$

c. $5\frac{2}{8}$
$-1\frac{7}{8}$

d. $12\frac{1}{5}$
$-3\frac{3}{5}$

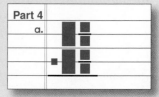

     Connecting Math Concepts

Lesson 91

Part 5 Figure out what each item equals.

a. $T(7 + 8 + 9)$ $\boxed{T = 2}$

b. $R(1 + 8 - 3)$ $\boxed{R = 7}$

c. $2(M \div 3)$ $\boxed{M = 30}$

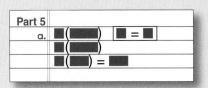

Part 6 Work each problem.

a. $\dfrac{5}{2} \div \dfrac{1}{10}$ b. $\dfrac{2}{9} \div 6$ c. $8 \div \dfrac{1}{4}$

Part 7 Find the volume of each cylinder. Start with the letter equation.

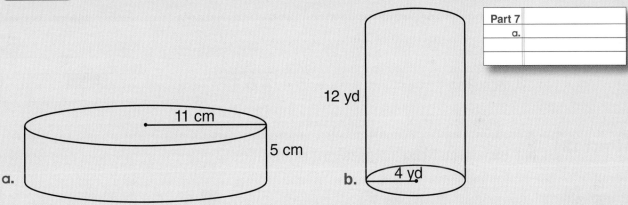

a. (11 cm, 5 cm)

b. (12 yd, 4 yd)

Part 8 Work each problem.

a. There were ducks and geese on the pond. The ratio of ducks to all birds was 5 to 8. There were 60 ducks. How many geese were there?

b. $\dfrac{2}{3}$ of the houses had garages. There were 120 houses. How many did not have garages? How many had garages?

Lesson 91

Part 9   Write each problem and work it.

a. Show the quantity $14 - 8 + 3$. Then multiply by 9.

b. Show the quantity $\frac{1}{3} + \frac{14}{3}$. Then multiply by 5.

c. Show the quantity $65 \div 5$. Then multiply by 11.

d. Show the quantity 2×26. Then multiply by $\frac{1}{4}$.

Part 10   Copy and complete each equation.

a. $.403 \times 10^2 = $ ▮

b. $\dfrac{.89}{10^2} = $ ▮

c. $\dfrac{360}{10^3} = $ ▮

d. $.02 \times 10^3 = $ ▮

Part 11   Copy and work each problem.

a. $39\overline{)140}$

b. $54\overline{)461}$

Part 12   Copy each item. Make the sign >, <, or =.

a. $\dfrac{4}{8}$ ▮ $\dfrac{1}{4}$

b. $\dfrac{1}{9}$ ▮ $\dfrac{6}{54}$

c. $\dfrac{8}{35}$ ▮ $\dfrac{1}{5}$

Connecting Math Concepts

Lesson 92

Part 1 Work each problem. Write the answer as a number and a unit name.

a. A bag has 14 pounds of sand in it. Sarah makes piles that are each $\frac{2}{7}$ pound. How many piles does she make?

b. Each cup holds $\frac{1}{5}$ quart. How many cups can be filled if there are 3 quarts of water?

c. A cook divided a salad into 4 equal servings. There were $\frac{7}{2}$ pounds of salad. How much did each serving weigh?

d. A pie weighs $\frac{8}{3}$ pounds. It is cut into slices that each weigh $\frac{1}{9}$ pound. How many slices are there?

Part 1	
a.	

Part 2

a. Change 3 years into seasons.

b. Change 7 minutes into seconds.

c. Change 9 weeks into days.

d. Change 6 feet into inches.

Facts:
- 1 foot is 12 inches.
- 1 week is 7 days.
- 1 year is 4 seasons.
- 1 minute is 60 seconds.

Part 2	
a.	

Part 3 Copy and work each problem. Below, write the statement with the sign >, <, or =.

a. $\frac{8}{2}\left(\frac{3}{3}\right) = \blacksquare$

b. $\frac{8}{3}\left(\frac{4}{3}\right) = \blacksquare$

c. $\frac{3}{5}\left(\frac{9}{10}\right) = \blacksquare$

d. $\frac{7}{4}\left(\frac{5}{8}\right) = \blacksquare$

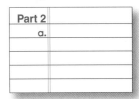

Part 4 Write the letter equation for each item.

a. $\frac{3}{5}$ of the trees were firs. There were 20 trees.

b. $\frac{7}{8}$ of the dogs were sleeping. 24 dogs were awake.

c. $\frac{2}{9}$ of the apples were ripe. There were 18 apples.

d. $\frac{5}{8}$ of the children were boys. There were 27 girls.

e. $\frac{3}{5}$ of the tigers were sleeping. There were 30 sleeping tigers.

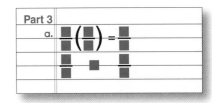

Lesson

Part 5 Copy and work each problem.

a. $30\frac{14}{15}$
$+ \ 5\frac{9}{15}$

b. $3\frac{5}{8}$
$+ 8\frac{5}{8}$

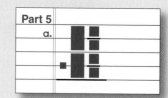

Part 6 Figure out what each item equals.

a. $P(1 - 1 + 10)$ $\boxed{P = 9}$

b. $4(12 - K)$ $\boxed{K = 6}$

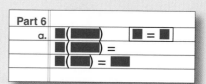

Part 7 Write each problem and work it.

a. Show the quantity $\frac{3}{4} \times 4$. Then multiply by $\frac{1}{2}$.

b. Show the quantity $20 \div 4$. Then multiply by $\frac{3}{2}$.

c. Show the quantity $56 - 21$. Then multiply by 6.

d. Show the quantity $\frac{1}{2} \times \frac{4}{5}$. Then multiply by 2.

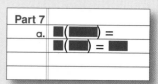

Part 8 Answer the question.

Monday	Tuesday	Wednesday	Thursday	Friday
9°	14°	4°	18°	0°

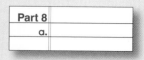

a. The table shows the high temperature for different days.

What is the average high temperature for the days shown?

Part 9 Write two equations for each item.

a. $\frac{5}{12}$ of the children were girls.

b. $\frac{2}{11}$ of the trees were dead.

c. $\frac{10}{13}$ of the windows were clean.

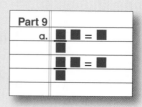

Lesson 92

Part 10 Copy and work each problem.

a. 14
   $- 5\frac{2}{7}$

b. $12\frac{2}{9}$
   $- 8\frac{5}{9}$

c. $6\frac{1}{8}$
   $- 4\frac{3}{8}$

d. $5\frac{2}{5}$
   $- 1\frac{4}{5}$

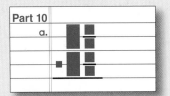

Part 11 Work each problem. Write the answer as a mixed number with a unit name.

a. The birch tree was $2\frac{4}{8}$ feet taller than the peach tree. The birch tree was $27\frac{5}{8}$ feet tall. How tall was the peach tree?

b. The truck started out with $26\frac{3}{4}$ tons of coal. The truck dropped off $15\frac{1}{4}$ tons of coal. How much coal was still on the truck?

c. Mr. Gray lost $14\frac{3}{5}$ pounds last summer and $4\frac{1}{5}$ pounds in the spring. How much weight did he lose in both seasons?

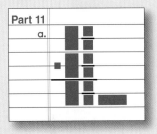

Lesson

Part 1 Copy and work each problem. Below, write the statement with the sign >, <, or =.

a. $\dfrac{3}{2}\left(\dfrac{5}{9}\right) = $ ■

b. $\dfrac{3}{8}\left(\dfrac{1}{2}\right) = $ ■

c. $\dfrac{2}{9}\left(\dfrac{3}{3}\right) = $ ■

d. $\dfrac{9}{5}\left(\dfrac{6}{4}\right) = $ ■

Part 1	
a.	■ (■) = ■
	■ ■ ■
	■ ■ ■
	■ ■ ■

Part 2 Find the volume of the rectangular prism in each position.

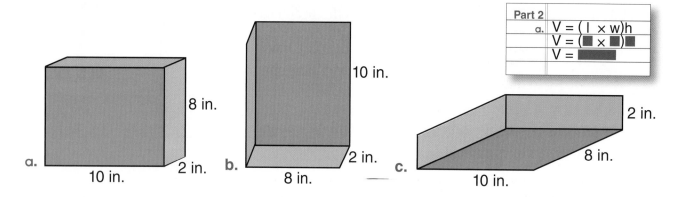

Part 2	
a.	V = (l × w)h
	V = (■ × ■) ■
	V = ■

a. 8 in. — 10 in. — 2 in.

b. 10 in. — 8 in. — 2 in.

c. 2 in. — 8 in. — 10 in.

Part 3 Work each problem.

a. $\dfrac{2}{9}$ of the flowers were not white. 54 flowers were not white. How many flowers were there in all? How many flowers were white?

Part 3	
a.	■ ■ _ ■
	■

b. $\dfrac{2}{9}$ of the windows were broken. 28 windows were not broken. How many windows were there? How many broken windows were there?

c. $\dfrac{5}{8}$ of the fish were large. There were 40 fish. How many small fish were there? How many large fish were there?

d. $\dfrac{3}{5}$ of the doors in a building were open. There were 60 closed doors. How many doors were in the building? How many doors were open?

Connecting Math Concepts

Lesson

Part 4 Copy and work each problem.

a. $12\frac{1}{4}$
$+18\frac{1}{5}$

b. $27\frac{5}{8}$
$-11\frac{1}{4}$

c. $38\frac{1}{6}$
$+ 2\frac{4}{9}$

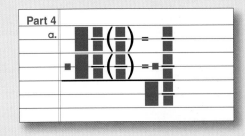

Part 5 Work each problem. Write the answer as a number and a unit name.

a. A factory uses $\frac{22}{3}$ pounds of iron to make 13 spikes. How much iron is used to make each spike?

Part 5	
a.	

b. 5 equal servings are made from a roast that weighs $\frac{13}{4}$ pounds. How much does each serving weigh?

Part 6 Answer the question.

Hours Billy Worked During the Weekend

Part 6	
a.	

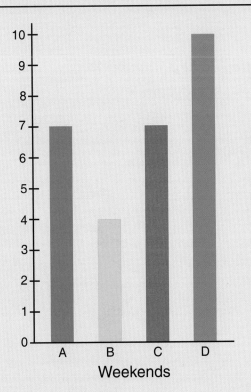

a. What is the average number of hours Billy worked on the weekends shown?

Lesson 93

Part 7   Copy and work each problem.

a. $4\frac{3}{8}$
    $+ \;\; \frac{5}{8}$

b. $2\frac{6}{10}$
    $+5\frac{9}{10}$

c. $20\frac{3}{4}$
    $+ \; 11\frac{7}{4}$

Part 7	
a.	

Part 8   Find the volume of each figure.

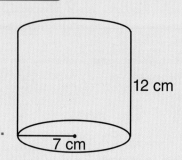

a.   12 cm   7 cm

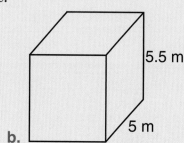

b.   5.5 m   5 m   5 m

Part 8	
a.	$V = (A_b)h$

Part 9   Copy and work each problem.

a. $\dfrac{4}{3} \div 7$ b. $\dfrac{9}{5} \div \dfrac{15}{2}$ c. $\dfrac{12}{7} \div \dfrac{2}{7}$ d. $\dfrac{11}{8} \div \dfrac{3}{4}$

Part 9	
a.	

Part 10   Copy and work each problem.

a. $8\overline{)672}$ b. $9\overline{)581}$ c. $7\overline{)503}$

d. $6\overline{)555}$ e. $7\overline{)290}$ f. $8\overline{)426}$

Part 10	
a.	

Part 11   Replace the letter with a number. Then work the problem.

a. $9(F - 3)$ $\boxed{F = 12}$ b. $M(12 \div 3)$ $\boxed{M = 6}$

Part 11	
a.	

c. $5(11 + B - 1)$ $\boxed{B = 6}$

Lesson

Part 1 Work each problem.

a. $\frac{2}{7}$ of the students were in the gym. 25 students were not in the gym. How many students were there? How many students were in the gym?

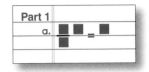

b. $\frac{1}{3}$ of the cookies were burned. There were 24 cookies in all. How many cookies were burned? How many cookies were not burned?

c. After the storm, $\frac{3}{11}$ of the windows were not broken. There were 18 windows that were not broken. How many windows were there in all? How many windows were broken?

d. $\frac{3}{8}$ of the horses were sleeping. 45 horses were awake. How many horses were sleeping? How many horses were there in all?

Part 2 Find the volume of the rectangular prism in each position.

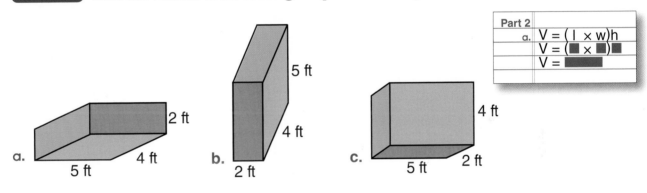

a. 5 ft 4 ft 2 ft

b. 5 ft 4 ft 2 ft

c. 4 ft 5 ft 2 ft

Part 3 Work each problem. Write the answer as a mixed number with a unit name.

a. Last week, Tom used up $3\frac{2}{5}$ gallons of gas. This week he used up $7\frac{4}{5}$ gallons. How many gallons did he use up during the two-week period?

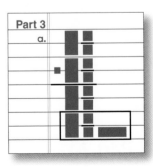

b. Juanita had $9\frac{1}{4}$ cups of flour. She used $4\frac{3}{4}$ cups making bread. How many cups of flour does she have left?

Connecting Math Concepts

Lesson 94

Part 4 Answer the question.

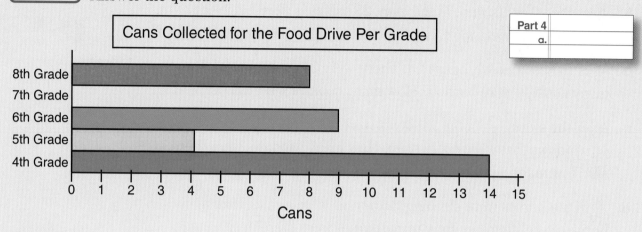

Cans Collected for the Food Drive Per Grade

Part 4	
a.	

a. What is the average number of cans collected today in 4th through 8th grades?

Part 5 Find the volume of each figure.

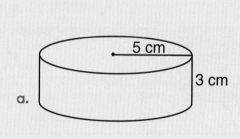 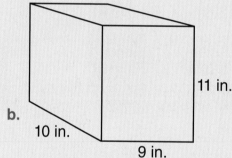

Part 5	
a.	

a.

b.

11 in.

10 in.

9 in.

Part 6 Copy and work each problem.

a. $14\frac{4}{5}$
$- \ 3\frac{1}{6}$

b. $3\frac{3}{8}$
$+ \ 11\frac{1}{3}$

Part 6	
a.	

Part 7 Write each problem. Then work it.

a. Show the quantity $38 \div 2$. Then multiply by $\frac{3}{2}$.

b. Add $12 + 18$. Then multiply 3.

Part 7	
a.	

Lesson  94

Part 8   Work each problem. Write the answer as a number and a unit name.

a. You want to make servings of ice cream that are $\frac{1}{3}$ pint each. You have 9 pints of ice cream. How many servings can be made?

b. 35 pounds of grapes are put in bags that each hold $\frac{1}{3}$ pound of grapes. How many bags are filled?

c. Ms. Harris has 9 ounces of silver. She puts an equal amount into 6 containers. How much silver is in each container?

Part 8	
a.	

Part 9   Copy and work each problem.

a. $3\frac{4}{5}$
$-1\frac{7}{5}$

b. 2
$-1\frac{3}{8}$

c. $27\frac{11}{9}$
$-\ 5\frac{4}{9}$

Part 9	
a.	

Part 10   Copy and work each problem.

a. $\frac{3}{4} \div \frac{4}{20}$

b. $5 \div \frac{2}{5}$

c. $\frac{1}{10} \div 8$

Part 10	
a.	

Lesson 95

Part 1 Use the facts to work each problem.

a. Change 300 minutes into hours.

b. Change 27 quarts into gallons.

c. Change 48 years into months.

d. Change 48 hours into days.

Facts:
- 1 year is 12 months.
- 1 gallon is 4 quarts.
- 1 day is 24 hours.
- 1 hour is 60 minutes.

Part 1	
a.	

Part 2 Figure out the volume of the stairs.

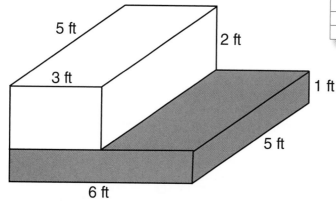

5 ft
2 ft
3 ft
1 ft
5 ft
6 ft

Part 2	
a.	$V = (A_b)h$ $V = (A_b)h$

Part 3 Work each problem. Write the answer as a mixed number with a unit name.

a. The school had $24\frac{1}{8}$ gallons of milk in the kitchen. The students drank $21\frac{6}{8}$ gallons of milk at lunch. How many gallons of milk were left?

b. On Wednesday, Shen ran $6\frac{4}{10}$ miles. On Saturday, Shen ran $11\frac{7}{10}$ miles. How many miles did he run in all?

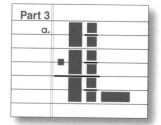

Independent Work

Part 4 Copy and work each problem.

a. $7\overline{)269}$ b. $8\overline{)364}$ c. $9\overline{)729}$

d. $7\overline{)450}$ e. $8\overline{)987}$ f. $9\overline{)789}$

Lesson 95

Part 5 Answer the question.

Centimeters of Rain that Fell on Different Days			
Monday	Tuesday	Wednesday	Thursday
5	10	2	8

a. What is the average rainfall for these days?

Part 5	
a.	

Part 6 Copy and work each problem.

a. $25\frac{6}{10}$
     $-19\frac{1}{2}$

b. $56\frac{1}{9}$
     $+12\frac{5}{6}$

Part 6	
a.	

Part 7 Work each problem.

a. In the factory, the ratio of new machines to old machines was 3 to 4. There were 63 machines in all. How many were new? How many were old?

b. $\frac{4}{5}$ of the students in the school spoke English. There were 400 students in the school. How many students did not speak English? How many students spoke English?

c. There were 8 broken chairs in the room. $\frac{9}{10}$ of the chairs were not broken. How many chairs were in the room? How many chairs were not broken?

Part 7	
a.	

Part 8 Copy and work each problem.

a. $\frac{4}{5}$
     $+6\frac{3}{5}$

b. $18\frac{1}{2}$
     $+ 1\frac{1}{2}$

c. $9\frac{14}{20}$
     $+3\frac{3}{20}$

Part 8	
a.	

Lesson 95

Part 9 Work each problem. Write the answer as a number and a unit name.

a. $\frac{36}{5}$ pounds of popcorn fill 9 bags. How many pounds are in each bag?

b. Fran cuts tape into pieces that are $\frac{3}{4}$ inch each. The tape is 12 inches long. How many pieces can she make?

c. The factory used $\frac{3}{2}$ ounces of latex to make 5 gloves. How much latex was used for each glove?

Part 9	
a.	

Part 10 Copy and work each problem. Below, write the statement with the sign >, <, or =.

a. $\frac{3}{5} \left(\frac{4}{3} \right) = $ ▮

b. $\frac{12}{7} \left(\frac{1}{2} \right) = $ ▮

c. $\frac{1}{5} \left(\frac{1}{7} \right) = $ ▮

Part 10	
a.	▮ (▮) = ▮
	▮ . ▮

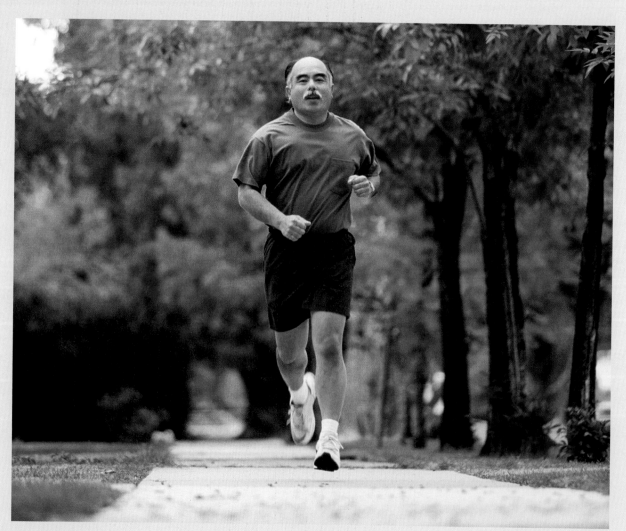

Connecting Math Concepts

Lesson 96

Part 1 — Work each item.

a. Change 5 years into months.

b. Change 12 days into hours.

c. Change 16 seasons into years.

d. Change 36 inches into feet.

e. Change 15 hours into minutes.

Facts:
- 1 hour is 60 minutes.
- 1 foot is 12 inches.
- 1 year is 12 months.
- 1 dollar is 20 nickels.
- 1 year is 4 seasons.
- 1 day is 24 hours.

Part 1	
a.	

Part 2 — Copy and work each problem.

a.
$$5\frac{4}{5}$$
$$+\,7\frac{3}{10}$$

b.
$$12\frac{5}{7}$$
$$+\,6\frac{1}{2}$$

c.
$$10\frac{3}{4}$$
$$+\,15\frac{5}{6}$$

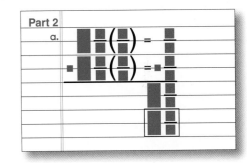

Part 3 — Figure out the volume of the stairs.

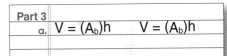

a.

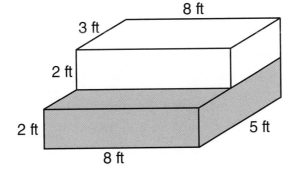

8 ft
3 ft
2 ft
2 ft
8 ft
5 ft

b.

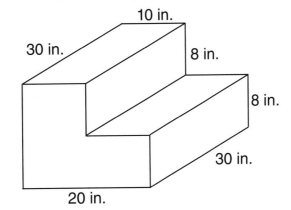

10 in.
30 in.
8 in.
8 in.
30 in.
20 in.

Lesson 96

Part 4 Figure out the average distance for the weights shown. Then answer the questions.

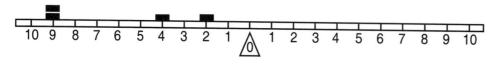

Part 4	
a.	

a. At what distance would you show the weights on the right side?

b. How many weights would you show on the right side?

c. What would be the total distance on the right side?

Independent Work

Part 5 Answer each question.

- You want to balance the beam by putting weights at one distance on the left side.

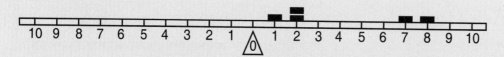

Part 5	
a.	

a. At what distance would you show the weights on the left side?

b. How many weights would you show on the left side?

c. What would be the total distance on the left side?

Part 6 Work each problem. Write the answer as a mixed number with a unit name.

a. The time required to make the old type of rack was $3\frac{4}{5}$ minutes. The time required to make the new type of rack is $4\frac{3}{5}$ minutes. How much more time is needed to make the new type of rack?

b. Each cinder block weighed $3\frac{5}{8}$ pounds more than the concrete block. The concrete block weighed $17\frac{3}{8}$ pounds. How much did the cinder block weigh?

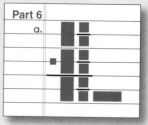

Part 7 Copy each problem. Replace the letter with a number. Then work the problem.

a. $P(35 \div 7)$ $P = 4$ b. $6(J + 20)$ $J = 4$

Part 7	
a.	

Lesson 96

Part 8 — Answer the question.

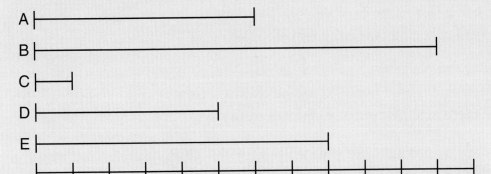

a. What is the average length of these lines?

Part 8	
a.	

Part 9 — Copy and work each problem.

a. $36\frac{7}{10}$
$-12\frac{1}{3}$

b. $4\frac{3}{5}$
$+18\frac{2}{8}$

Part 9	
a.	

Part 10 — Copy and work each problem.

a. $\frac{4}{3} \div \frac{1}{9}$

b. $1 \div \frac{3}{5}$

c. $\frac{9}{10} \div 2$

Part 10	
a.	

Part 11 Work each problem.

a. There were 2 white onions for every 3 red onions. There were 96 white onions. How many onions were there in all? How many were red?

Part 11	
a.	

b. $\frac{7}{9}$ of the cherries were ripe. There were 182 unripe cherries. How many cherries were there in all? How many cherries were ripe?

c. $\frac{2}{3}$ of the students could ice skate. 48 students could ice skate. How many students could not ice skate? How many students were there?

Part 12 Copy and work each problem.

a.
$$35\frac{4}{5}$$
$$-29\frac{3}{5}$$

b.
$$78$$
$$-51\frac{1}{8}$$

c.
$$29\frac{1}{4}$$
$$-7\frac{3}{4}$$

Part 12	
a.	

Lesson 97

Part 1
Write the probability equation for pulling a white marble from each can.

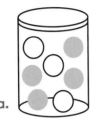 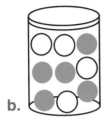 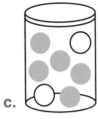 

a.   b.   c.   d.

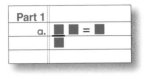

Part 1	
a.	■ ■ = ■
	■

Part 2
Work each item.

a. Change 200 nickels into dollars.

b. Change 8 years into months.

c. Change 30 hours into days.

d. Change 55 minutes into hours.

e. Change 14 feet into inches.

Facts:
- 1 hour is 60 minutes.
- 1 foot is 12 inches.
- 1 year is 12 months.
- 1 dollar is 20 nickels.
- 1 year is 4 seasons.
- 1 day is 24 hours.

Part 2	
a.	

Part 3
Copy and work each problem.

a.
$$9\frac{5}{8}$$
$$+7\frac{5}{6}$$

b.
$$10\frac{4}{5}$$
$$+4\frac{7}{20}$$

c.
$$8\frac{3}{9}$$
$$+6\frac{2}{6}$$

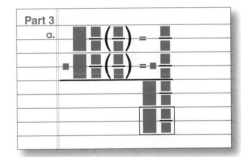

Part 4 Figure out the volume of the stairs.

a.

20 in.

5 in.

10 in.

5 in.

10 in.

30 in.

Part 4		
a.	$V = (A_b)h$	$V = (A_b)h$

b.

3 ft

.6 ft

1 ft

.6 ft

3 ft

2 ft

Independent Work

Part 5 Work each problem. Write the answer as a mixed number with a unit name.

a. The pine posts weighed $1\frac{2}{3}$ pounds less than the oak posts. The oak posts weighed $18\frac{1}{3}$ pounds. How much did the pine posts weigh?

b. A bucket of salt water weighed $1\frac{5}{8}$ pounds more than a bucket of fresh water. The bucket of fresh water weighed $16\frac{3}{8}$ pounds. How much did a bucket of salt water weigh?

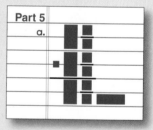

Part 6 Copy and work each problem. Below, write the statement with the sign >, <, or =.

a. $\frac{1}{7}\left(\frac{3}{4}\right) = \blacksquare$

b. $\frac{4}{9}\left(\frac{5}{5}\right) = \blacksquare$

c. $\frac{4}{5}\left(\frac{1}{9}\right) = \blacksquare$

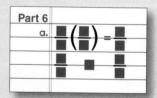

Part 7 Write each problem. Then work it.

a. Show the quantity $64 + 4 - 30$. Then multiply by $\frac{1}{4}$.

b. Show the quantity $56 - 12$. Then multiply by 2.

Connecting Math Concepts

Lesson 97

Part 8 Work each problem. Write the answer as a number and a unit name.

a. How many $\frac{3}{4}$-pound servings can be made from 48 pounds of salad?

b. The factory makes 18 hangers from $\frac{9}{2}$ pounds of plastic. How much does each hanger weigh?

Part 8	
a.	

Part 9 Copy and work each problem.

a. $33\frac{1}{3}$
   $+ \ 6\frac{3}{5}$

b. $29\frac{4}{9}$
   $- 20\frac{1}{3}$

c. $37\frac{5}{8}$
   $+ 47\frac{2}{6}$

Part 9	
a.	

Part 10 Work each problem.

a. $\frac{1}{6}$ of the cakes were in the oven. 60 cakes were not in the oven. How many cakes were there? How many cakes were in the oven?

Part 10	
a.	

b. 300 adults were at a concert. The ratio of adults to men was 5 to 2. How many men were at the concert? How many women were at the concert?

c. In a cafeteria, $\frac{3}{7}$ of the plates were being used. 84 plates were not being used. How many plates were there? How many plates were being used?

Part 11 Answer each question.

- You want to balance the beam by placing weights at one distance on the right side.

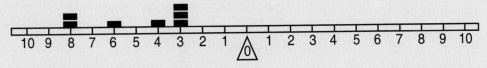

Part 11	
a.	

a. At what distance would you show the weights on the right side?

b. How many weights would you show on the right side?

c. What would be the total distance on the right side?

Lesson 98

Part 1
Work each item. Use the measurement table if you need to.

a. Change 6 pints into cups.

b. Change 3 years into weeks.

c. Change 3 days into weeks.

d. Change 120 months into years.

e. Change 2 days into hours.

f. Change 252 inches into yards.

Part 1	
a.	

Part 2
Write the equation for each color.

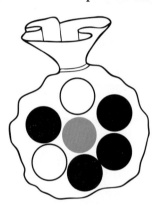

a. white

black

red

Part 2		
a. ■ ■ = ■ ■	b. ■ ■ = ■
	■ ■ = ■
■ ■ = ■ ■	
■ ■ = ■ ■	

Part 3
Figure out the volume of the stairs.

a.

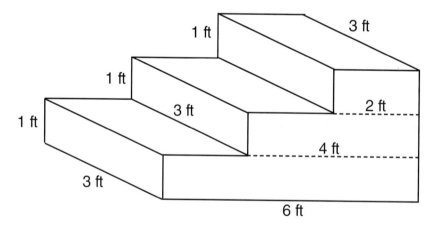

Part 3	
a.	$V = (A_b)h$

Lesson

Part 4 Work each problem.

a. $\frac{3}{4}$ of the birds were flying. 60 birds were not flying. How many birds were there in all? How many birds were flying?

Part 4	
a.	

b. The ratio of cabins to passengers is 2 to 3. There are 36 cabins. How many passengers are there?

c. $\frac{5}{8}$ of the fish were at the east end of the lake. There were 400 fish at the east end. How many fish were in the lake? How many were not at the east end?

Part 5 Copy each problem and multiply. Below, write the statement with the sign >, <, or =.

a. $\frac{7}{3}\left(\frac{5}{6}\right)$ b. $\frac{1}{10}\left(\frac{9}{10}\right)$ c. $\frac{23}{13}\left(\frac{3}{10}\right)$

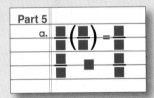

Part 6 Answer each question.

- You want to balance the beam by placing weights at one distance on the left side.

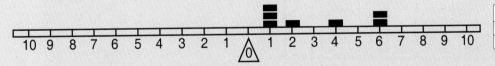

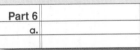

a. At what distance would you show the weights on the left side?

b. How many weights would you show on the left side?

c. What would be the total distance on the left side?

Part 7 Copy each problem. Replace the letter with a number. Then work the problem.

a. $7(64 - K)$ $\boxed{K = 10}$ b. $4(B + 15)$ $\boxed{B = 11}$

Part 7	
a.	

Lesson 98

Part 8
Copy and work each problem.

a. $5\frac{3}{4}$
$-1\frac{1}{8}$

b. $3\frac{4}{6}$
$+10\frac{7}{9}$

c. $1\frac{7}{9}$
$-\frac{3}{4}$

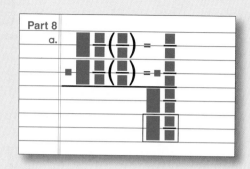

Part 9
Work each problem. Write the answer as a mixed number with a unit name.

a. The truck dumped $2\frac{3}{5}$ tons of dirt in the front yard and $4\frac{4}{5}$ in the back yard. How many tons of dirt did the truck dump?

Part 9	
a.	

b. In February and March, the children in Lowell School ate $23\frac{1}{5}$ pounds of peanut butter. They ate $17\frac{3}{5}$ pounds in March. How many pounds did they eat in February?

Part 10
Copy and work each problem.

a. $\frac{7}{8} \div \frac{3}{4}$

b. $5 \div \frac{10}{7}$

c. $\frac{9}{14} \div 3$

Part 10	
a.	

Connecting Math Concepts

Lesson

Part 1 Copy each mixed number. Figure out the fraction it equals.

a. $7\dfrac{3}{8}$ b. $1\dfrac{6}{11}$ c. $9\dfrac{4}{5}$ d. $4\dfrac{2}{9}$

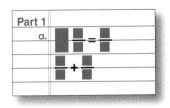

Part 2 Work each item. Use the measurement table if you need to.

a. Change 30 minutes into hours.

b. Change 7 pounds into ounces.

c. Change 690 feet into yards.

d. Change 3 miles into yards.

e. Change 62 days into weeks.

f. Change 4 gallons into cups.

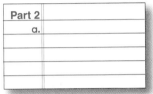

Part 3 Work each item.

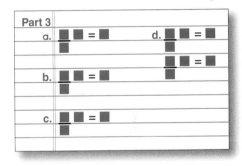

Write the equation for:

a. green

b. purple

c. yellow

d. If you took trials until you pulled out 9 green marbles, about how many trials would you expect to take?

e. If you took 96 trials, about how many yellow marbles would you expect to pull out?

f. If you took 16 trials, about how many purple marbles would you expect to pull out?

g. If you took trials until you pulled out 10 purple marbles, about how many trials would you expect to take?

Lesson 99

Independent Work

Part 4 Write each problem. Then work it.

a. Show the quantity 360 – 90 multiplied by $\frac{1}{3}$.

b. Show the quantity 45 – 15 + 6 multiplied by $\frac{3}{4}$.

Part 4	
a.	

Part 5 Copy and work each problem.

a. $6\overline{\smash{)}535}$　　b. $9\overline{\smash{)}218}$　　c. $7\overline{\smash{)}345}$　　d. $4\overline{\smash{)}695}$

Part 5	
a.	■ ▬

Part 6 Answer each question.

- You want to balance the beam by placing weights at one distance on the left side.

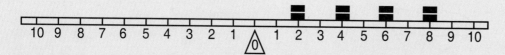

Part 6	
a.	

a. At what distance would you show the weights on the left side?

b. How many weights would you show on the left side?

c. What would be the total distance on the left side?

Part 7 Find the volume of each figure.

Part 7	
a.	

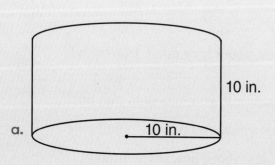

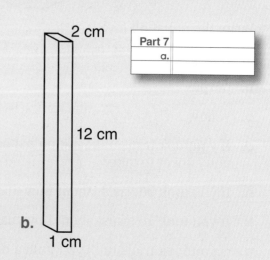

Connecting Math Concepts

Lesson 99

Part 8 Copy and work each problem.

a. $7\frac{1}{3}$
   $+8\frac{8}{9}$

b. $12\frac{2}{3}$
   $-\ 3\frac{1}{5}$

c. $4\frac{3}{4}$
   $+10\frac{7}{8}$

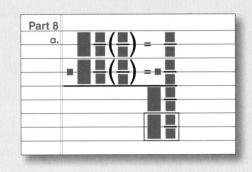

Part 9 Write each problem in a column and work it.

a. $3.07 + 354.1 + 26$

b. $2.07 - 1.006$

c. $.002 + 1.006 + .453$

d. $2.904 + 6.009 + 14.05$

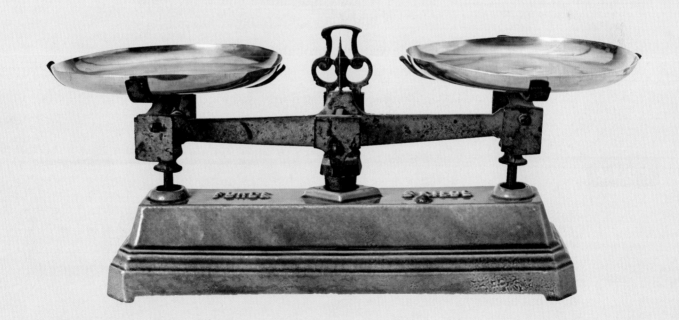

Lesson 100

Part 1 Work each item. Use the measurement table if you need to.

a. Change 300 seconds into minutes.

b. Change 660 years into decades.

c. Change 4 gallons into cups.

d. Change 62 weeks into days.

e. Change 45 ounces into pounds.

Part 1	
a.	

Part 2 Work each item.

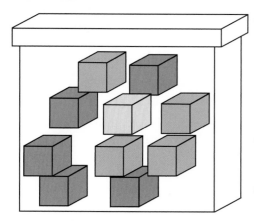

a. Write the probability equation for pulling a blue cube from the box without looking.

b. If you took 25 trials, about how many blue cubes would you expect to pull?

c. Write the probability equation for the yellow cube.

d. If you took trials until you pulled out 7 yellow cubes, about how many trials would you expect to take?

e. Write the probability equation for red cubes.

f. If you took 18 trials, about how many red cubes would you expect to pull out?

Part 2	
a.	■ ■ = ■ ■

Part 3

a. 5 out of every 9 envelopes are sealed.

b. 9 out of every 15 cars were moving.

c. 2 out of every 3 people were visitors.

d. 7 out of every 10 cats are sleeping.

Part 3	
a.	■ ■ = ■ ■

Lesson 100

Independent Work

Part 4 Copy each mixed number. Figure out the fraction it equals.

a. $3\frac{2}{7}$

b. $10\frac{1}{4}$

c. $6\frac{3}{5}$

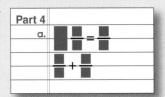

Part 5 Answer the question.

Pounds of Apples Used at the Juice Shop on Different Days

Part 5	
a.	

a. What is the average number of pounds used on these days?

Lesson 100

Part 6 Work each problem.

a. There were 5 wet days for every 2 sunny days. There were 56 days in all. How many were sunny? How many were wet?

b. $\frac{2}{7}$ of the birds were white. 55 birds were not white. How many birds were there in all? How many birds were white?

c. $\frac{7}{8}$ of the workers liked popcorn. There were 224 workers. How many did not like popcorn? How many liked popcorn?

Part 6	
a.	

Part 7 Copy and work each problem. Below, write the statement with the sign >, <, or =.

a. $\frac{10}{9}\left(\frac{7}{7}\right) = \blacksquare$ b. $\frac{4}{5}\left(\frac{4}{7}\right) = \blacksquare$ c. $\frac{1}{9}\left(\frac{8}{3}\right) = \blacksquare$

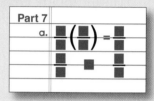

Part 8 Copy and work each problem.

a. $13\frac{3}{8}$ b. $3\frac{5}{9}$ c. $4\frac{3}{5}$ d. 9 e. $11\frac{3}{7}$
$+13\frac{7}{8}$ $+4\frac{4}{9}$ $+6\frac{1}{5}$ $-\frac{3}{8}$ $-5\frac{4}{7}$

Part 8	
a.	

Part 9 Answer each question.

- You want to balance the beam by placing weights at one distance on the right side.

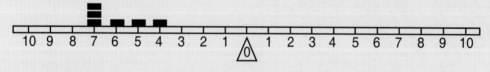

Part 9	
a.	

a. At what distance would you show the weights on the right side?

b. How many weights would you show on the right side?

c. What would be the total distance on the right side?

Lesson 100

Part 10 Copy the problem. Replace the letter with a number. Then work the problem.

a. $3(5 + D)$ $\boxed{D = 12}$ b. $A(64 \div 2)$ $\boxed{A = 3}$

Part 10	
a.	

Part 11 Work each problem. Write the answer as a mixed number with a unit name.

a. The yard had $46\frac{1}{3}$ pounds of leaves on the ground. Henry and Nan removed $24\frac{2}{3}$ pounds. How many pounds of leaves were still on the ground?

Part 11	
a.	

b. The washer weighed $73\frac{2}{5}$ pounds. The drier weighed $68\frac{3}{5}$ pounds. What was the weight of both appliances?

c. Tom was $62\frac{3}{8}$ inches tall. His dad was $74\frac{5}{8}$ inches tall. How much shorter was Tom than his dad?

Lesson 101

Part 1 Copy each problem. Simplify. Then multiply.

a. $\dfrac{50}{100}\left(\dfrac{20}{3}\right) =$

b. $\dfrac{35}{100}\left(\dfrac{10}{28}\right) =$

c. $\dfrac{20}{50}\left(\dfrac{10}{3}\right) =$

Part 2 Copy and work each problem.

a. $39\overline{)1\,0\,5\,7}$

b. $91\overline{)4\,9\,2\,4}$

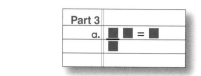

Part 3 Work each item.

a. How many pink rabbits would you expect to pull out if you took 12 trials?

b. About how many trials would you expect to take to pull 4 blue rabbits from the hat?

c. How many yellow rabbits would you expect to pull out if you took 30 trials?

Part 4 Work each problem. Write the answer as a mixed number with a unit name.

a. The cow gained $16\frac{1}{2}$ pounds last week and $14\frac{7}{8}$ pounds this week. How much weight did the cow gain during the two-week period?

b. Last fall, the maple tree was $12\frac{1}{4}$ feet tall. This fall, the tree is $15\frac{5}{6}$ feet tall. How much did the tree grow during the year?

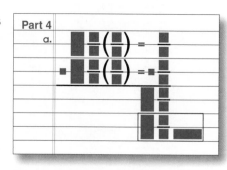

Lesson 101

Independent Work

Part 5

Work each problem. The measurement fact table is on the inside back cover.

a. Change 16 weeks into days.

b. Change 15 feet into yards.

c. Change 3 days into hours.

d. Change 28 months into years.

Part 5	
a.	

Part 6

Copy and complete each equation.

a. $\frac{.15}{10^3}$ = ■ b. 2.03×10^2 = ■ c. $\frac{.06}{10^2}$ = ■

Part 6	
a.	$\frac{■}{■}$ = ■

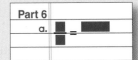

Part 7

Write each problem in a column and work it.

a. 23.05 – 1.992

b. 4.503 + .045 + 2.882

c. .234 + 6.543 + .08

Part 7	
a.	

Part 8

Copy and work each problem.

a. $8\overline{)6\,6\,5}$ b. $6\overline{)1\,9\,7}$ c. $7\overline{)5\,8\,5}$ d. $8\overline{)9\,3\,7}$

Part 8	
a.	$■\overline{)■}$

Part 9

Write each problem. Then work it.

a. Add 64 + 64. Then multiply by $\frac{3}{8}$.

b. Show the quantity $\frac{4}{5} + \frac{8}{5}$ multiplied by 10.

Part 9	
a.	

Part 10

Find the volume of each figure.

Part 10	
a.	

a.
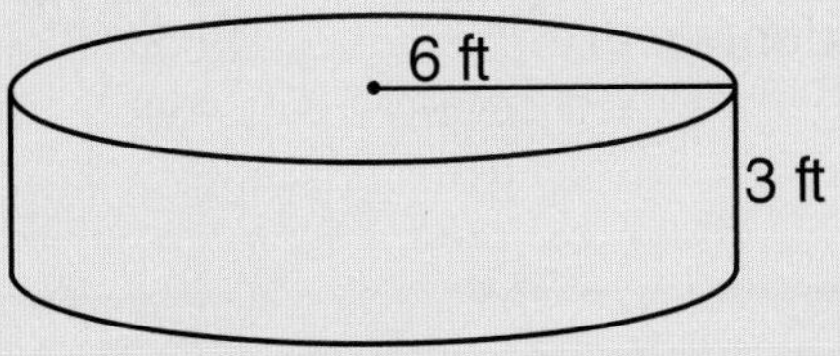

b.
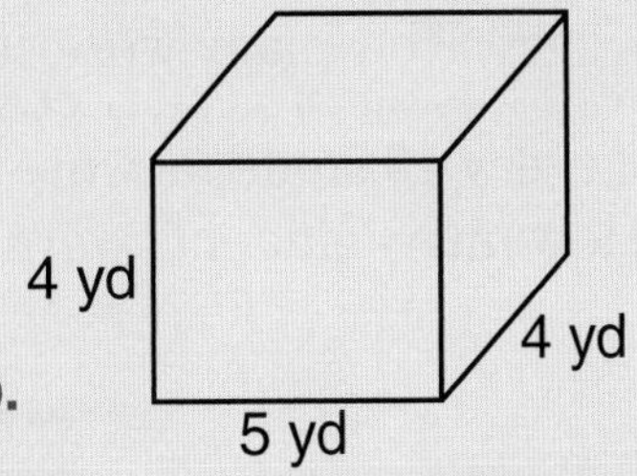

Lesson 102

Part 1
Copy and work each problem.

a. $18\overline{)679}$ b. $62\overline{)1736}$

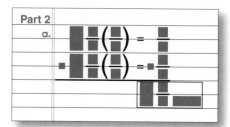

Part 2
Work each problem. Write the answer as a mixed number with a unit name.

a. Tony finished the canoe race in $4\frac{1}{3}$ days. Henry finished the race in $6\frac{3}{4}$ days. How many days faster was Tony than Henry?

b. A fuel tank had $15\frac{5}{8}$ gallons of gas in it. $7\frac{9}{16}$ more gallons were poured into the tank. How many gallons ended up in the tank?

Part 3
Write each decimal value.

a. One hundred thirteen thousandths

b. Eleven and five tenths

c. Sixty and ninety-nine hundredths

d. One and seven thousandths

e. Forty-five hundredths

Part 4
Work each problem.

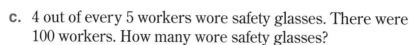

a. 5 out of every 7 packages were frozen. There were 35 frozen packages. How many packages were there in all?

b. 2 out of every 9 bugs were spiders. There were 54 bugs. How many spiders were there?

c. 4 out of every 5 workers wore safety glasses. There were 100 workers. How many wore safety glasses?

d. 3 out of every 10 students wore jeans. 36 students wore jeans. How many students were there?

Connecting Math Concepts

Lesson 102

Independent Work

Part 5 Copy each problem. Simplify. Then multiply.

a. $\dfrac{45}{100}\left(\dfrac{50}{9}\right) =$

b. $\dfrac{24}{80}\left(\dfrac{30}{21}\right) =$

c. $\dfrac{25}{100}\left(\dfrac{8}{10}\right) =$

d. $\dfrac{7}{4}\left(\dfrac{16}{28}\right) =$

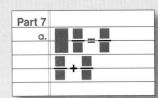

Part 6 Work each problem. The measurement fact table is on the inside back cover of your textbook.

a. Change 11 pounds into ounces.

b. Change 3 years into months.

c. Change 7 yards into feet.

d. Change 5 days into weeks.

Part 7 Write an equation to show the fraction that equals each mixed number.

a. $2\dfrac{5}{8}$

b. $3\dfrac{5}{9}$

c. $2\dfrac{2}{5}$

d. $4\dfrac{3}{8}$

Part 8 Work each problem.

a. On Monday, Sid rode his bike 7 miles. On Wednesday, he rode 5 miles. On Friday, he rode 11 miles. What was the average number of miles he rode on the three days?

b. Over the weekend students read from a book about China. Jan read 17 pages of the book. Bonnie read 51 pages. Tina read 40 pages. What was the average number of pages the students read?

Part 9 Copy and work each problem.

a. $72\overline{)7223}$

b. $68\overline{)1435}$

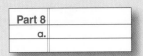

Lesson 102

Part 10 Work the item.

a. Find the volume of concrete needed to build the staircase.

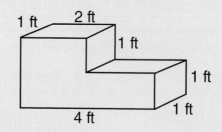

Part 10	
a.	

Part 11 Work each problem.

a. The ratio of zebras to animals was 2 to 9. There were 720 animals in all. How many were not zebras? How many were zebras?

Part 11	
a.	

b. The mixture contained iron and copper. The mixture weighed 875 pounds. $\frac{4}{5}$ of the weight was iron. What was the weight of the iron? What was the weight of the copper?

Part 12 Copy each problem and multiply. Below, write the statement with the sign >, <, or =.

a. $\frac{2}{3} \left(\frac{1}{5} \right)$

b. $\frac{3}{7} \left(\frac{9}{8} \right)$

c. $5 \left(\frac{11}{11} \right)$

d. $\frac{20}{13} \left(\frac{4}{3} \right)$

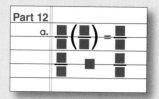

Lesson 103

Part 1 Work each item.

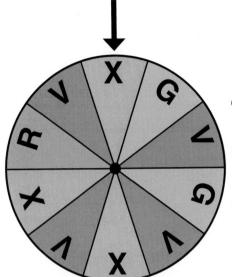

Part 1
a. ■ ■ = ■
     ■

a. If you spin the wheel 20 times, what's the expected number for the spinner stopping at X?

b. How many times would you expect to spin the wheel until it stops at V two times?

c. If you spin the wheel 40 times, what's the expected number for the spinner stopping at R?

d. How many times would you expect to spin the wheel until it stops at G three times?

Part 2 Write each decimal value.

a. Fourteen and eight thousandths

b. Five and forty-nine hundredths

c. Twenty-six thousandths

d. Forty and five tenths

e. One and three hundred twelve thousandths

Part 2
a.

Part 3 Write the new decimal value.

a. .725 (Multiply by 100.)

b. 115.8 (Divide by 10.)

c. 7.025 (Multiply by 1000.)

d. 52.18 (Multiply by 10.)

e. .031 (Multiply by 100.)

f. 35.4 (Divide by 100.)

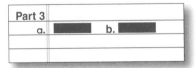

Independent Work

Part 4 Copy and work each problem.

a. 5⟌491   b. 7⟌491   c. 9⟌488   d. 6⟌593

Part 4
a. ■ ■

Lesson 103

Part 5 Work each item.

a. In a stone collection, 2 of every 10 stones were green. There were 760 stones. How many were green? How many were not green?

Part 5	
a.	

b. 3 of every 4 students had a perfect paper. There were 15 students who had perfect papers. How many students were there in all? How many did not have perfect papers?

Part 6 Find the area and perimeter of each figure.

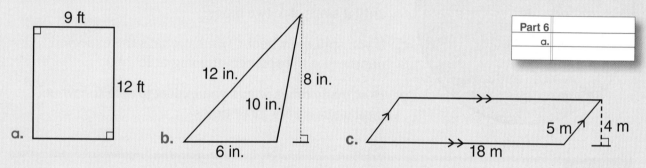

Part 6	
a.	

Part 7 Copy each problem. Simplify. Then multiply.

a. $\dfrac{60}{30}\left(\dfrac{20}{5}\right)$

b. $\dfrac{48}{80}\left(\dfrac{90}{60}\right)$

c. $\dfrac{100}{50}\left(\dfrac{300}{50}\right)$

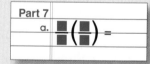

Part 8 Work each problem.

a. The trip from Boston to Miami took $3\dfrac{3}{5}$ hours. The trip from Boston to Chicago took $2\dfrac{4}{5}$ hours. How much shorter was the trip to Chicago than the trip to Miami?

b. The basket of apples weighed $28\dfrac{1}{6}$ pounds. The basket of plums weighed $37\dfrac{1}{3}$ pounds. How much heavier was the basket of plums than the basket of apples?

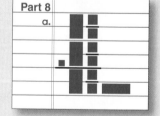

Part 9 Copy and work each problem.

a. $30\overline{)1529}$

b. $51\overline{)569}$

Part 9	
a.	

Lesson 103

Part 10
Copy and simplify each fraction.

a. $\dfrac{81}{36}$ b. $\dfrac{40}{75}$ c. $\dfrac{27}{99}$ d. $\dfrac{56}{28}$

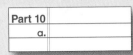

Part 11
Answer each question.

- You want to make the beam balance by putting weights at one place on the right side.

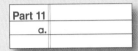

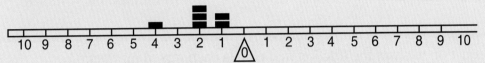

    a. How many weights would go on the right side?

    b. At what distance would you put the weights?

- You want to make the beam balance by putting weights at one place on the left side.

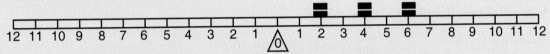

    c. How many weights would go on the left side?

    d. At what distance would you put the weights?

Part 12
Answer each question.

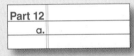

> There are 24 marbles in a bag. 8 are red, and the rest are black.

a. How many black marbles are there?

b. If you took 63 trials of pulling marbles from the bag, about how many would you expect to be black marbles?

c. About how many would you expect to be red marbles?

d. If you took 300 trials, about how many black marbles would you expect to pull from the bag?

Part 13
Copy and work each problem. Below, write the statement with the sign >, <, or =.

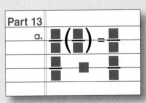

a. $\dfrac{9}{5}\left(\dfrac{6}{7}\right)$ b. $\dfrac{4}{3}\left(\dfrac{5}{3}\right)$ c. $\dfrac{4}{5}\left(\dfrac{7}{7}\right)$

Lesson 104

Part 1 — Write each decimal number.

a. Three and five hundredths

b. Three and five tenths

c. Three hundred four and five tenths

d. Thirteen thousandths

e. Fourteen hundredths

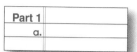

Part 1	
a.	

Part 2 — Write an equation to show the fraction that equals each mixed number.

a. $2\frac{5}{6} = \blacksquare$

b. $5\frac{1}{8} = \blacksquare$

c. $4\frac{4}{5} = \blacksquare$

d. $3\frac{2}{3} = \blacksquare$

e. $10\frac{1}{8} = \blacksquare$

Part 2	
a.	$\blacksquare \frac{\blacksquare}{\blacksquare} = \frac{\blacksquare}{\blacksquare}$

Part 3 — Write the new value.

a. 5.2 (Multiply by 100.)

b. .3 (Divide by 100.)

c. .78 (Divide by 10.)

d. .78 (Multiply by 1000.)

e. 8.34 (Multiply by 10.)

f. 8.34 (Divide by 100.)

Part 3		
a.		d.
b.		e.
c.		f.

Independent Work

Part 4 — Copy each fraction. Write the simplified value it equals.

a. $\frac{21}{7}$

b. $\frac{10}{40}$

c. $\frac{4}{20}$

d. $\frac{33}{6}$

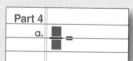

Part 4	
a.	$\frac{\blacksquare}{\blacksquare} =$

Part 5 — Find the area of each circle.

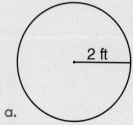

2 ft

a.

12 mi

b.

Part 5	
a.	

Lesson 104

Part 6 Work each problem. The measurement fact table is on the inside back cover.

   a. Change 4 years into months.

   b. Change 12 feet into yards.

   c. Change 4 days into hours.

   d. Change 4 months into years.

Part 6	
a.	

Part 7 Figure out each shaded angle.

Part 7	
a.	

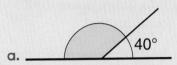

a. 40°

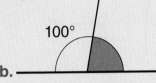

b. 100°

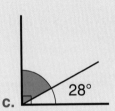

c. 28°

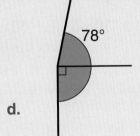

d. 78°

Part 8 Copy and work each problem.

   a. $70\overline{)5821}$ **b.** $21\overline{)1768}$

Part 8	
a.	

Part 9 Work each problem.

a. The bigger mouse weighed $1\frac{1}{2}$ ounces more than the smaller mouse. The bigger mouse weighed $9\frac{2}{3}$ ounces. How much did the smaller mouse weigh?

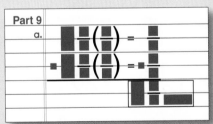

b. The maple tree was $13\frac{2}{5}$ feet taller than the pine tree. The pine tree was $77\frac{1}{2}$ feet tall. How tall was the maple tree?

Part 10 Copy and simplify each fraction.

   a. $\dfrac{3 \times 18}{4 \times 5}$ **b.** $\dfrac{12 \times 14}{7 \times 8}$ **c.** $\dfrac{3 \times 20}{9 \times 12}$

Part 10	
a.	

Lesson 105

Part 1
Answer each question.

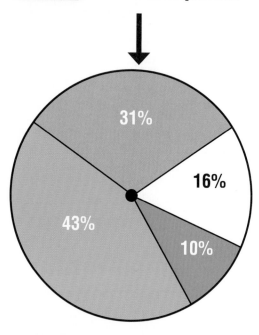

31%

16%

43%

10%

a. If you take 100 trials, about how many times would you expect the spinner to stop at white?

b. If you take 100 trials, about how many times would you expect the spinner to stop at brown?

c. If you take 100 trials, about how many times would you expect the spinner to stop at green?

d. If you take 100 trials, about how many times would you expect the spinner to stop at red?

Part 2
Work each problem.

> **Sample Problem**
> 6(25)
> 6(20) + 6(5)
> 120 + 30 = $\boxed{150}$

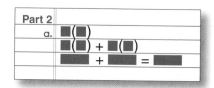

a. 9(42)   b. 2(57)   c. 7(83)

Part 3
Write an equation to show the fraction that equals each mixed number.

a. $1\frac{3}{7}$ b. $20\frac{1}{2}$ c. $3\frac{4}{5}$ d. $8\frac{1}{3}$ e. $4\frac{7}{10}$

Part 4
Write the new value.

a. .13   (Divide by 100.)

b. 2.03   (Divide by 10.)

c. 9.9   (Divide by 100.)

d. .099   (Multiply by 1000.)

e. 48.2   (Divide by 100.)

f. .102   (Multiply by 100.)

Lesson 105

Part 5 Work each item.

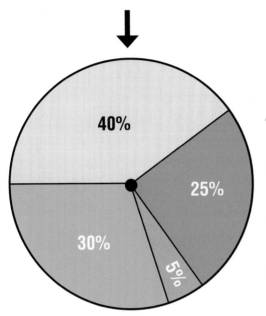

40%

25%

30%

5%

a. If you spin the wheel 10 times, what's the expected number for the wheel stopping at yellow?

b. You spin the wheel until it lands on purple 4 times. How many trials would you expect to take?

c. If you spin the wheel 30 times, what's the expected number for the wheel stopping at green?

d. If you spin the wheel 12 times, what's the expected number for the wheel stopping at red?

Independent Work

Part 6 Write each decimal number.

a. Seven and nine thousandths

b. Forty-eight hundredths

c. One and fifty-six thousandths

d. One hundred and one hundredth

e. Forty-five thousandths

f. Seven and three tenths

Part 6	
a.	

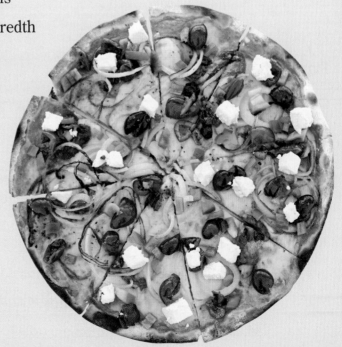

Lesson 105

Part 7 Answer each question.

- You want to make the beam balance by putting weights at one place on the right side.

Part 7	
a.	

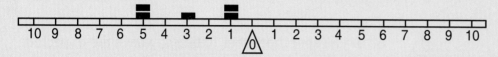

    a. How many weights would go on the right side?

    b. At what distance would you put the weights?

- You want to make the beam balance by putting weights at one place on the left side.

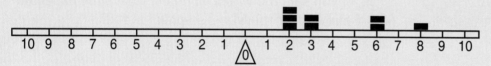

    c. How many weights would go on the left side?

    d. At what distance would you put the weights?

Part 8 Copy each problem. Simplify. Then multiply.

a. $\dfrac{40}{20}\left(\dfrac{20}{10}\right)$ b. $\dfrac{18}{12}\left(\dfrac{8}{2}\right)$ c. $\dfrac{36}{8}\left(\dfrac{48}{6}\right)$

Part 8	
a.	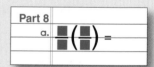

Part 9 Copy and work each problem.

    a. $36\overline{)428}$ b. $90\overline{)6721}$

Part 9	
a.	

Part 10 Work each problem. The measurement fact table is on the inside back cover.

    a. Change 6 days into weeks.

    b. Change 3 days into hours.

    c. Change 3 hours into minutes.

    d. Change 25 pounds into ounces.

Part 10	
a.	

Lesson 105

Part 11 Copy and work each problem. Below, write the statement with the sign >, <, or =.

a. $\dfrac{14}{15}\left(\dfrac{2}{1}\right)$ b. $\dfrac{12}{7}\left(\dfrac{4}{5}\right)$ c. $\dfrac{3}{19}\left(\dfrac{11}{10}\right)$

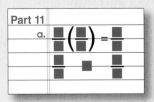

Part 12 Work each problem.

a. The truck started out with $7\dfrac{3}{5}$ tons of sand. The truck dumped $2\dfrac{4}{5}$ tons of sand. What was the weight of sand after the dump?

b. The plastic chair weighed $2\dfrac{1}{4}$ pounds less than the wooden chair. The plastic chair weighed $9\dfrac{3}{4}$ pounds. What was the weight of the wooden chair?

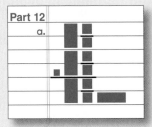

Part 13 Copy and work each problem.

a. $8\overline{)287}$ b. $4\overline{)631}$ c. $5\overline{)238}$ d. $3\overline{)502}$

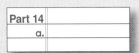

Part 14 Work each problem.

a. In a warehouse, there were raincoats and umbrellas. There were 5 umbrellas for every 3 raincoats. There were 1536 objects in the warehouse. How many were umbrellas? How many were raincoats?

b. In a pond, there were sunfish and bass. $\dfrac{6}{10}$ of the fish were sunfish. There were 400 bass. How many fish were there in all? How many sunfish were there?

c. There were 120 workers cleaning the park. 7 out of every 8 workers wore uniforms. How many workers did not wear uniforms? How many workers wore uniforms?

Lesson 106

Part 1
Write an equation to show the fraction that equals each mixed number.

a. $3\frac{2}{9}$
b. $5\frac{1}{7}$
c. $3\frac{3}{4}$
d. $6\frac{2}{5}$

Part 2
Use the facts to work each problem.

a. Change 43.7 meters into centimeters.

b. Change .7 kilometers into meters.

c. Change 456 meters into kilometers.

d. Change .004 centimeters into millimeters.

e. Change .8 centimeters into meters.

Facts:
- 1 kilometer is 1000 meters.
- 1 meter is 100 centimeters.
- 1 centimeter is 10 millimeters.

Independent Work

Part 3
Work each item.

a. Write the probability fraction for the spinner stopping on green.

b. Write the probability fraction for the spinner stopping on yellow.

- You spin the wheel 100 times.

c. About how many times would you expect the spinner to land on white?

d. About how many times would you expect the spinner to land on green?

e. About how many times would you expect the spinner to land on red?

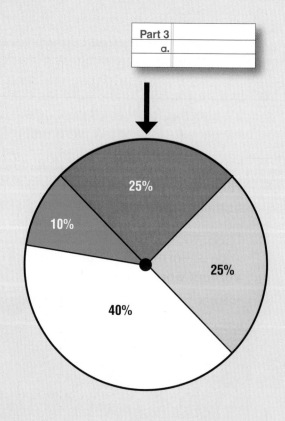

Lesson 106

Part 4 Find the area and perimeter of each figure.

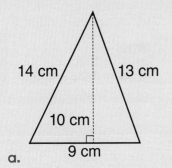

a. 14 cm, 13 cm, 10 cm, 9 cm

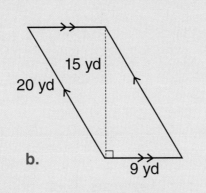

b. 15 yd, 20 yd, 9 yd

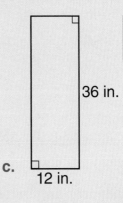

c. 36 in., 12 in.

Part 4	
a.	

Part 5 Work each item.

a. $\frac{3}{4}$ of the trees in a grove were birch. 120 trees in the grove were not birch. How many trees were in the grove? How many birch trees were in the grove?

Part 5	
a.	

b. 3 of every 7 students entered the contest. There were 57 students who entered the contest. How many students were there? How many students did not enter the contest?

Part 6 Write the answer to each problem.

a. .234 (Divide by 100.)

b. 3.06 (Multiply by 100.)

c. .0084 (Divide by 1000.)

d. 4.01 (Divide by 10.)

e. .703 (Multiply by 1000.)

f. 3.106 (Multiply by 10.)

Part 6	
a.	

Part 7 Copy each problem. Simplify. Then multiply.

a. $\frac{9}{100}\left(\frac{200}{18}\right)$

b. $\frac{16}{10}\left(\frac{50}{8}\right)$

c. $\frac{300}{45}\left(\frac{9}{60}\right)$

Part 7	
a.	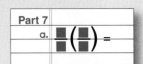

Part 8 Copy and work each problem.

a. $52\overline{)2080}$

b. $31\overline{)2561}$

Part 8	
a.	

Lesson 106

Part 9 Copy each problem. Write the multiplication for the tens and ones. Then write the addition equation.

a. 5(44)   b. 6(56)   c. 9(21)   d. 3(42)

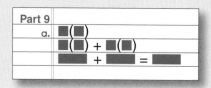

Part 10 Work each problem.

a. The average Coldport refrigerator lasts $21\frac{3}{12}$ years. The average Everware refrigerator lasts $40\frac{1}{3}$ years. How much longer do the Everware refrigerators last?

b. There were dogs and cats in the shelter. The average weight of the dogs was $29\frac{4}{5}$ pounds. The average weight of the cats was $8\frac{2}{10}$ pounds. How much heavier was the average dog than the average cat?

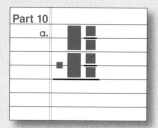

Lesson 107

Independent Work

Part 1 Complete each equation to show the fraction that equals each mixed number.

a. $5\frac{1}{7}$ b. $3\frac{4}{5}$ c. $4\frac{3}{7}$ d. $7\frac{9}{10}$ e. $4\frac{3}{5}$

Part 1	
a.	▮▮ = ▮

Part 2 Write each decimal number.

a. Fourteen hundredths

b. Fourteen and six hundredths

c. Eleven and seventy-six hundredths

d. Three hundred four thousandths

e. One and eighteen thousandths

f. Four tenths

g. Sixteen thousandths

Part 2	
a.	

Part 3 Work each item.

• You take 50 trials at spinning the wheel.

Part 3	
a.	

a. Write the probability equation for landing on white.

b. About how many times would you expect to land on white?

c. Write the probability equation for landing on blue.

d. About how many times would you expect to land on blue?

e. Write the probability equation for landing on green.

f. About how many times would you expect to land on green?

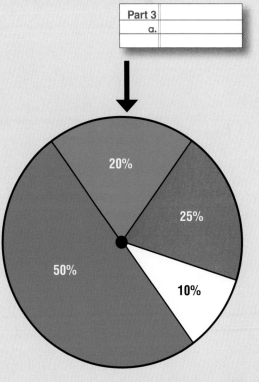

Lesson 107

Part 4 Copy each problem. Simplify. Then multiply.

a. $\dfrac{70}{90}\left(\dfrac{5}{21}\right)$

b. $\dfrac{45}{130}\left(\dfrac{260}{5}\right)$

c. $\dfrac{400}{3}\left(\dfrac{12}{200}\right)$

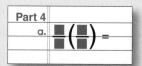

Part 5 Work each problem.

a. The workers mixed gravel and cement to make concrete. The gravel in the mixture weighed $2\frac{1}{4}$ tons. The cement in the mixture weighed $10\frac{5}{8}$ tons. How much more was the weight of the cement than the weight of the gravel?

b. Geneva was $3\frac{3}{8}$ years older than Tonia. Geneva was $10\frac{3}{4}$ years old. How old was Tonia?

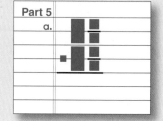

Part 6 Copy and work each problem.

a. $63\overline{)2039}$

b. $43\overline{)698}$

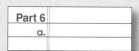

Part 7 Copy each problem. Write the multiplication for the tens and ones. Then write the addition equation.

a. 7(25)

b. 9(51)

c. 3(86)

d. 5(77)

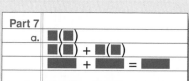

Part 8 Copy and work each problem. Below write the statement with the sign >, <, or =.

a. $\dfrac{12}{12}\left(\dfrac{7}{5}\right)$

b. $\dfrac{4}{5}\left(\dfrac{11}{11}\right)$

c. $15\left(\dfrac{9}{8}\right)$

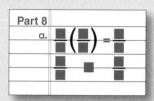

Connecting Math Concepts

Lesson 107

Part 9 Work each item.

a. During a camping trip, on $\frac{1}{4}$ of the days it rained. The trip lasted 88 days. On how many days did it rain? On how many days did it not rain?

Part 9	
a.	

b. In a gem collection, 5 out of every 11 gems was a ruby. There were 150 rubies in the collection. How many gems were not rubies? How many gems were there in all?

Part 10 Copy and work each problem.

a. $6\overline{)559}$ b. $7\overline{)838}$ c. $9\overline{)308}$ d. $8\overline{)519}$

Part 10	
a.	

Lesson 108

Part 1 Work each problem.

a. A rope was 90 feet long. 15 yards were cut from the rope. How many feet long is the rope now?

b. The carpenter had 3 boards. One was 3 yards long. One was 45 inches long. The last board was 2 yards long. What's the total number of yards for all three boards?

c. There were 27 gallons of milk in the tank. Then 11 quarts of milk were added. How many gallons of milk are in the tank now?

Part 1	
a.	

Part 2 Work each problem. Use the measurement table at the back of this book if you need to.

a. Change .5 centimeters into meters.

b. Change 228 milligrams into grams.

c. Change 15 liters into centiliters.

d. Change 65 centigrams into grams.

e. Change .67 grams into milligrams.

Part 2	
a.	

Independent Work

Part 3 Work each problem.

a. The rabbit weighed $17\frac{2}{3}$ ounces more than the squirrel. The squirrel weighed $24\frac{1}{2}$ ounces. How much did the rabbit weigh?

b. The average boy in the class read $2\frac{3}{4}$ books during the year. The average girl read $3\frac{1}{4}$ books during the year. How many more books did the average girl read than the average boy read?

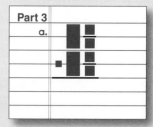

Part 4 Write the answer to each problem.

a. 42.3 (Divide by 100.)

b. .034 (Multiply by 1000.)

c. 3.56 (Multiply by 10.)

Part 4	
a.	

Lesson 108

Part 5   Copy each mixed number. Write the fraction it equals.

a. $4\dfrac{3}{9} =$ ▮   b. $12\dfrac{1}{3} =$ ▮   c. $7\dfrac{5}{9} =$ ▮

Part 5	
a.	

Part 6   Find the average for the line plot. Show your answer as a whole number or a mixed number.

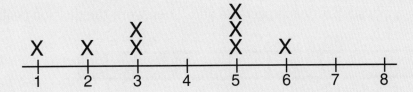

Part 6	

Part 7   Work each item. Use the measurement table if you need to.

    a. Change 4 pounds into ounces.

    b. Change 19 inches into feet.

    c. Change 10 feet into yards.

    d. Change 7 yards into feet.

Part 7	
a.	

Part 8   Work each item.

   • There are 12 marbles in a bag. 4 are red. 6 are blue.

Part 8	
a.	

a. If you took 48 trials of pulling marbles from the bag, about how many blue marbles would you expect to pull out?

b. About how many trials would you expect to take to pull out 10 red marbles?

Lesson 109

Part 1
Copy each problem. Write the new problem below. Simplify. Then multiply.

a. $2\frac{3}{10} \times \frac{20}{7}$
b. $\frac{9}{5} \times 6\frac{2}{3}$
c. $\frac{1}{7} \times 2\frac{4}{5}$
d. $3\frac{3}{8} \times \frac{4}{9}$

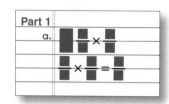

Part 2
Write the multiplication for each part and add. Then work the division problem.

a.

2 | 30 | 4

b.

5 | 20 | 3

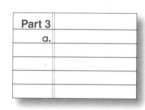

Part 3
Work each problem.

a. A dog weighed 46 pounds. A cat weighed 96 ounces. How many pounds did the two animals weigh together?

b. The loaf of bread lasted 3 days. The mustard lasted 12 weeks. How many more days did the mustard last than the bread?

c. The cat was 39 months old. The dog was 3 years old. How many months older was the cat than the dog?

Part 3
a.

Part 4
Find the surface area of each rectangular prism.

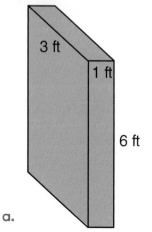

3 ft
1 ft
6 ft

a.

Part 4
a. SA = 2(■ + ■ + ■)

10 in.
2 in.
5 in.

b.

Lesson 109

Part 5 — Copy and work each problem.

a. $5\frac{1}{5}$
$+3\frac{2}{3}$

b. $4\frac{4}{9}$
$-4\frac{1}{3}$

c. $8\frac{5}{8}$
$+6\frac{3}{4}$

Part 5	
a.	

Part 6 — Write the answer to each problem.

a. 400.3 (Multiply by 100.)

b. 40.7 (Divide by 100.)

c. 3.14 (Multiply by 1000.)

Part 6	
a.	

Part 7 — Work each problem. Write the answer as a mixed number with a unit name.

a. The old truck weighs $5\frac{6}{10}$ tons. The new truck weighs $4\frac{1}{4}$ tons. How much do the two trucks weigh together?

b. Greg read $3\frac{2}{5}$ books last year. He read $6\frac{7}{8}$ books this year. How many books did he read during both years?

Part 7	
a.	

Part 8 — Copy each mixed number. Write the fraction it equals.

a. $6\frac{3}{5} = $ ■

b. $8\frac{4}{9} = $ ■

c. $9\frac{3}{10} = $ ■

Part 8	
a.	

Part 9 — Write each answer.

a. Change 3.046 centimeters into meters.

b. Change .4 kilometers into meters.

c. Change 45.203 centiliters into milliliters.

Part 9	
a.	

Lesson 109

Part 10 — Work each item.

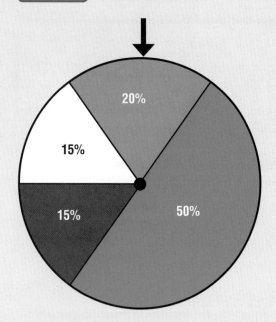

20%

15%

15%

50%

Part 10	
a.	

a. If you took 100 trials of spinning the spinner, about how times would you expect to land on red?

b. About how many times would you expect to land on black?

c. About how many times would you expect to spin to land on black 3 times?

d. About how many times would you expect to spin to land on white 6 times?

Part 11 — Work each item. Use the measurement table if you need to.

a. Change 5 ounces into pounds.

b. Change 3 pounds into ounces.

c. Change 4 feet into yards.

d. Change 10 yards into feet.

Part 11	
a.	

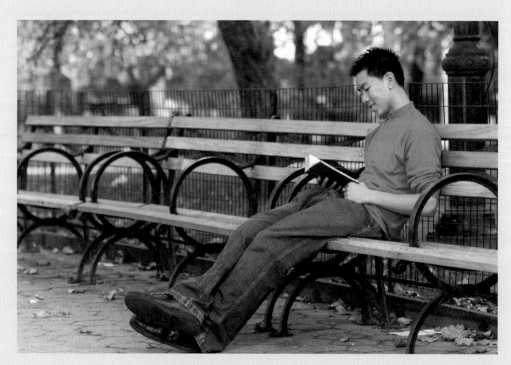

Connecting Math Concepts

Lesson 110

Part 1 Write the names for the figures in a through e. Then answer the questions.

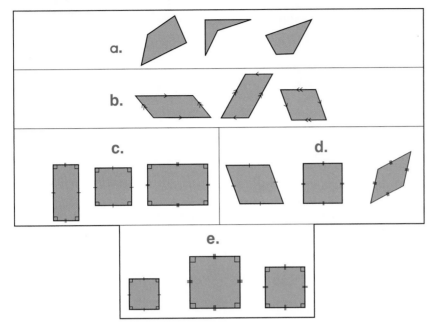

Part 1	
a.	

f. What is any rectangle with all sides the same length?

g. What is any quadilateral with two pairs of parallel sides?

h. What is any parallelogram with four 90-degree angles?

i. What is any parallelogram with all sides the same length?

Part 2 Write the multiplication for each part and add. Then use the diagram to work the division problem.

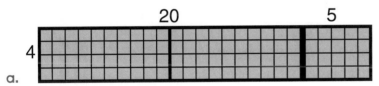

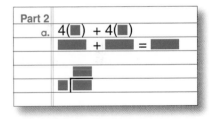

Part 3 Work each problem.

a. Tank X held 12 gallons of oil. Tank Y held 30 pints of oil. How many more pints were there in Tank X than in Tank Y?

b. Driveway T was 69 feet long. Driveway R was 12 yards long. How many yards longer was driveway T than driveway R?

c. Cindy went hiking for 13 days in May and 2 weeks in September. How many weeks did she hike in all?

Part 3	
a.	

Lesson 110

Part 4 Find the surface area of the rectangular prism.

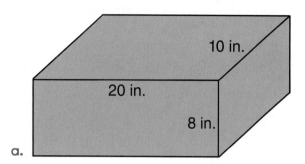

a.

Part 4	
a.	SA = 2(■ + ■ + ■)

Independent Work

Part 5 Copy and work each problem.

a. $1\frac{2}{5} \times \frac{3}{4}$

b. $\frac{5}{8} \times 1\frac{3}{10}$

c. $\frac{6}{5} \times 3\frac{1}{2}$

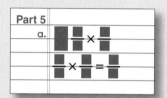

Part 6 Work each problem.

a. The oldest player on the baseball team was $17\frac{1}{3}$ years older than the youngest player on the team. The youngest player was $19\frac{3}{4}$ years old. How old was the oldest player on the team?

b. The trip from Los Angeles to New York took $6\frac{1}{3}$ hours. The trip from Los Angeles to Chicago took $4\frac{2}{3}$ hours. How much longer was the trip to New York than the trip to Chicago?

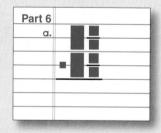

Part 7 Copy each mixed number. Write the fraction it equals.

a. $3\frac{6}{11} = $ ■

b. $5\frac{4}{7} = $ ■

c. $8\frac{7}{10} = $ ■

Part 7	
a.	

Part 8 Find the average. Your answer will be a mixed number.

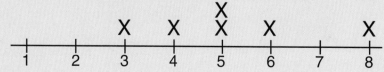

Part 8	

Connecting Math Concepts

Lesson 110

Part 9 Work each item. Use the measurement table if you need to.

  a. Change 3 weeks into days.

  b. Change 11 days into weeks.

  c. Change 20 ounces into pounds.

  d. Change 20 pounds into ounces.

Part 9	
a.	

Part 10 Answer each question.

- You want to make the beam balance by putting weights at one place on the right side.

Part 10	
a.	

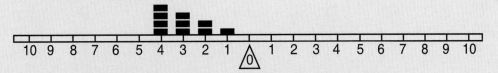

  a. How many weights would go on the right side?

  b. Where would the weights go on the right side?

Part 11 Work each item.

  - There are 16 marbles in a bag. 2 are yellow. 10 are red.

Part 11	
a.	

a. If you took 32 trials of pulling marbles from the bag, about how many yellow marbles would you expect to pull out?

b. About how many trials would you expect to take to pull out 20 yellow marbles?

Part 12 Write each answer. Use the measurement table if you need to.

  a. Change 500 milligrams into grams.

  b. Change 3.62 meters into centimeters.

  c. Change .034 kilometers into meters.

Part 12	
a.	

Lesson 111

Part 1 These line plots show the length in feet of different boards. Find the average length of the boards.

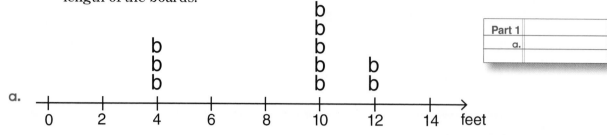

Part 1	
a.	

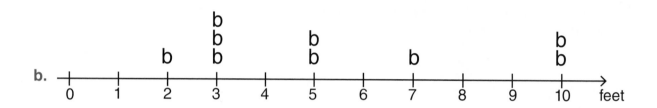

Part 2 Write the names for the figures in a through e. Then answer the questions.

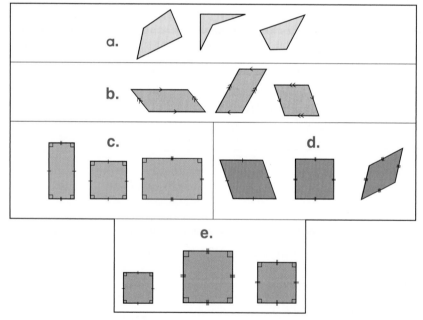

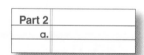

Part 2	
a.	

f. What is any quadrilateral that has two pairs of parallel sides?

g. What is any figure that has four straight sides and four angles?

h. What is any parallelogram that has four 90-degree angles and four sides the same length?

i. What is any parallelogram that has four sides the same length?

j. What is any parallelogram that has four 90-degree angles?

Lesson 111

Part 3 Write the multiplication for each part and add. Then work the division problem.

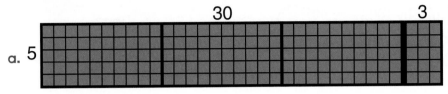

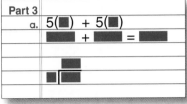

Part 3	
a.	5(■) + 5(■)
	■ + ■ = ■
	■
	■⟌■

Part 4 Write the letter equation for each sentence.

a. The truck was $\frac{9}{2}$ the weight of the car.

b. The wallet has $\frac{3}{4}$ as much money as the piggy bank.

c. The dinner costs $\frac{7}{4}$ as much as the show.

d. The rainfall in August was $\frac{9}{5}$ as much as the rainfall in July.

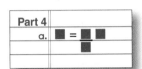

Part 4	
a.	■ = ■ ■
	■

Part 5 Find the surface area of the rectangular prism.

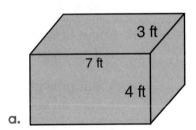

3 ft

7 ft

4 ft

a.

Part 5	
a.	SA = 2(■ + ■ + ■)

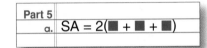

Independent Work

Part 6 Work each problem.

a. Tim is 70 inches tall. Roberto is 6 feet tall. How many inches taller is Roberto than Tim?

b. Boat A weighed 2.1 tons. Boat B weighed 1700 pounds. How many pounds do the boats weigh together?

c. A tree was 28 inches tall. It grew 3 feet in the next five years. How many feet tall was the tree then?

Part 6	
a.	

Lesson 111

Part 7
Write the answer to each problem.

    a. 74.32   (Multiply by 100.)

    b. .13   (Divide by 10.)

    c. 53.4   (Divide by 1000.)

Part 7	
a.	

Part 8
Work the problem. Write the answer as a mixed number with a unit name.

a. This morning Rob picked $8\frac{2}{3}$ baskets of corn. This afternoon he picked $5\frac{1}{2}$ baskets of corn. How many more baskets of corn did he pick this morning than he picked this afternoon?

Part 8	
a.	

Part 9
Work each item.

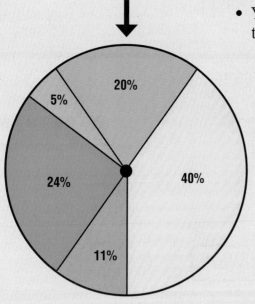

Part 9	
a.	

• You take 25 trials of spinning the spinner.

    a. About how many times would you expect to land on red?

    b. About how many times would you expect to land on green?

    c. About how many times would you expect to land on yellow?

Part 10
Copy each mixed number. Write the fraction it equals.

a. $4\frac{7}{9} =$ ▮   b. $2\frac{5}{8} =$ ▮   c. $6\frac{6}{7} =$ ▮

Part 10	
a.	

Part 11
Work each item. Use the measurement table if you need to.

a. Change 4 feet into inches.   b. Change 4 inches into feet.

c. Change 3 weeks into days.   d. Change 13 days into weeks.

Part 11	
a.	

Lesson 112

Part 1
Write the names for the figures in a through e. Then answer the questions.

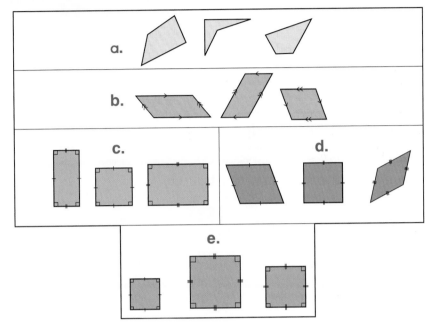

Part 1	
a.	

f. What is any parallelogram that has four 90-degree angles?

g. What is any figure that has four straight sides and four angles?

h. What is any parallelogram that has four 90-degree angles and four sides the same length?

i. What is any parallelogram that has four sides the same length?

j. What is any quadrilateral that has two pairs of parallel sides?

Part 2
Work each problem.

a. A pipe leaked at the rate of $1\frac{2}{5}$ gallons each day. The pipe leaked for 8 days. How many gallons did the pipe leak?

b. A recipe calls for $1\frac{1}{2}$ cups of oil. How much oil is needed for $\frac{3}{4}$ of the amount the recipe calls for?

c. Two students wrote pages of notes. Jay wrote $4\frac{3}{4}$ pages. Fran wrote $1\frac{3}{5}$ times as much as Jay. How many pages did Fran write?

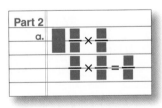

Lesson 112

Part 3   Work each problem.

a. The camera weighed $\frac{3}{5}$ as much as the book. The camera weighed $\frac{9}{4}$ pounds. How many pounds did the book weigh?

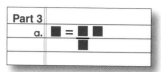

b. Tamir picked $\frac{3}{4}$ as many apples as his mom picked. His mom picked 12 apples. How many apples did Tamir pick?

Independent Work

Part 4   Copy each mixed number. Write the fraction it equals.

a. $9\frac{5}{7} = \blacksquare$ b. $8\frac{4}{10} = \blacksquare$ c. $7\frac{8}{9} = \blacksquare$

Part 4	
a.	

Part 5   Write each answer.

a. Change 4.6 meters into centimeters.

b. Change 276.4 milligrams into grams.

c. Change 3.3 liters into milliliters.

Part 5	
a.	

Part 6   Work each item.

a. What's $\frac{4}{3}$ of $2\frac{1}{2}$ days?

b. What's $\frac{7}{8}$ of 5 pounds?

c. What's $\frac{3}{5}$ of $\frac{1}{4}$ year?

Part 6	
a.	

Part 7   Find the average.

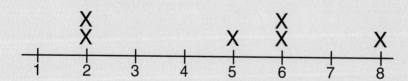

Part 7	

Lesson 112

Part 8 Find the surface area of each rectangular prism.

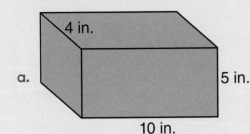

4 in.

5 in.

10 in.

a.

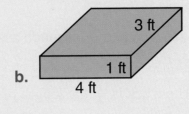

3 ft

1 ft

4 ft

b.

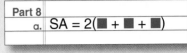

Part 8	
a.	SA = 2(■ + ■ + ■)

Part 9 Work each item.

a. A ditch was 10 yards long. Workers made it 40 feet longer. How many yards long is it now?

b. A rabbit weighed 48 ounces. Then the rabbit gained 3 pounds. How many ounces does the rabbit weigh now?

c. A hole was 2 feet deep. Sam made it 48 inches deeper. How many feet deep is the hole now?

Part 9	
a.	

Part 10 Copy and work each problem.

a. $\frac{3}{4} \times 2\frac{1}{3}$

b. $\frac{4}{7} \times \frac{3}{8}$

c. $\frac{2}{5} \times \frac{5}{3}$

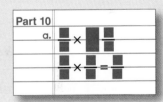

Part 10	
a.	

Part 11 Work each problem.

• There are 20 marbles in a bag. 2 are yellow. 10 are black.

a. About how many trials would you expect to take to pull out 4 yellow marbles?

b. If you took 50 trials, about how many black marbles would you expect to pull from the bag?

Part 11	
a.	

Part 12 Write the answer to each problem.

a. 53.7 (Divide by 1000.)

b. .035 (Multiply by 100.)

c. 6.45 (Divide by 10.)

Part 12	
a.	

Lesson 113

Part 1   Work each problem.

a. Jan weighed 120 pounds. Jan weighed $\frac{4}{5}$ as much as her mother. How much did her mother weigh?

b. The boat is 28 feet long. The truck is $\frac{7}{4}$ the length of the boat. How long is the truck?

c. Gross Hill is $\frac{3}{8}$ the height of Dillard Hill. Dillard Hill is 240 feet high. How high is Gross Hill?

Part 2   Find the average for each line plot.

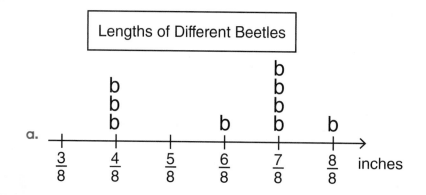

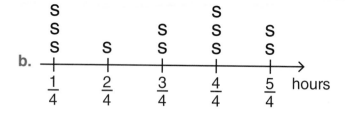

Part 3   Work each problem.

a. Jaylen collected $5\frac{2}{3}$ pounds of shrimp. Andrea collected 3 times as many pounds of shrimp. How many pounds did Andrea collect?

b. Ali's stand sold $8\frac{3}{4}$ gallons of orange juice this morning. Jan's stand sold $\frac{3}{5}$ as much orange juice. How much did Jan sell this morning?

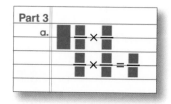

Connecting Math Concepts

Lesson 113

Part 4 | Write the answer to each problem.

a. .502 (Divide by 100.)

b. .003 (Multiply by 1000.)

c. 60.08 (Multiply by 100.)

d. .056 (Divide by 100.)

Part 4	
a.	

Part 5 | Copy each problem. Simplify. Then multiply.

a. $\dfrac{5}{15}\left(\dfrac{20}{60}\right)$ b. $\dfrac{12}{180}\left(\dfrac{9}{36}\right)$ c. $\dfrac{40}{15}\left(\dfrac{9}{10}\right)$

Part 5	
a.	

Part 6 | Work each problem.

a. 4 out of every 5 rings were made of gold. There were 100 rings. How many rings were made of gold? How many were not made of gold?

b. Last February in Rock City, 3 of every 7 days were snowy. It snowed on 12 days. On how many days did it not snow? How many fewer snow days were there than days that had no snow?

Part 6	
a.	

Part 7 | Write the multiplication for the tens and ones. Then write the addition equation.

a. 4(58)   b. 9(76)   c. 5(18)

Part 7	
a.	■(■)
	■(■) + ■(■)
	■■■ + ■■■ = ■■■

Part 8 | Work each item.

a. What's $\dfrac{4}{3}$ of $2\dfrac{1}{6}$ years?

b. What's $\dfrac{5}{6}$ of $7\dfrac{1}{2}$ pounds?

c. What's $\dfrac{4}{9}$ of $3\dfrac{1}{4}$ hours?

d. What's $\dfrac{9}{10}$ of $4\dfrac{1}{2}$ days?

Part 8	
a.	

Lesson 113

Part 9
Work the problem. Write the answer as a mixed number with a unit name.

a. Last week, the Wilson family drank $11\frac{1}{6}$ quarts of milk. This week the Wilson family drank $13\frac{3}{4}$ quarts of milk. What's the total amount of milk they drank during both weeks?

Part 9	
a.	

Part 10
Work each item. Use the measurement table if you need to.

a. Change .4920 centiliters into liters.

b. Change 11 meters into millimeters.

c. Change 14.72 grams into kilograms.

Part 10	
a.	

Part 11
Write an equation for each sentence.

a. The boat was $\frac{7}{8}$ the length of the car.

b. Silver costs $\frac{2}{19}$ as much as gold.

c. Henry made $\frac{5}{8}$ as many baskets as Jay made.

Part 11	
a.	

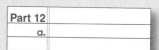

Part 12
Work each item.

a. The cook added 5 quarts of broth to 1 gallon of water. How many gallons was the mixture?

Part 12	
a.	

b. In the morning, Bonnie worked $3\frac{1}{2}$ hours. In the afternoon, she worked for 45 minutes. How many hours did she work that day?

Lesson 114

Part 1 Work each problem.

a. The amount they spent on dinner was $\frac{7}{4}$ the amount they spent on breakfast. They spent $56 on dinner. How much did they spend on breakfast?

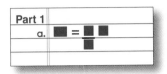

b. There are 36 students in the library. The number of students in the computer room is $\frac{3}{4}$ the number of students in the library. How many students are in the computer room?

c. The boat weighed $\frac{3}{7}$ as much as the truck. The truck weighed 14 tons. How much did the boat weigh?

d. Blue Lake has 480 fish. Blue Lake has $\frac{1}{5}$ as many fish as Cloud Lake. How many fish are in Cloud Lake?

Part 2 Find the average for each line plot.

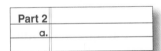

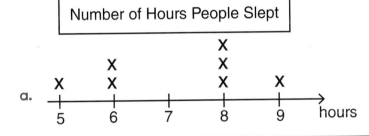

a. Number of Hours People Slept

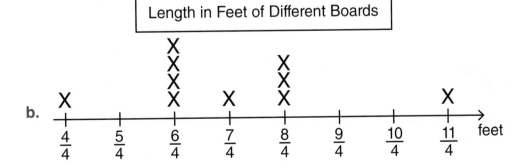

b. Length in Feet of Different Boards

Lesson 114

Part 3 Write the equation for each item.

    a. You divide 5 into 6 equal parts.

    b. You divide 7 into 6 equal parts.

    c. You divide 6 into 2 equal parts.

    d. You divide 12 into 5 equal parts.

    e. You divide 12 into 3 equal parts.

    f. You divide 12 into 17 equal parts.

Part 3	
a.	■ ÷ ■ = ■

Part 4 Work each problem.

a. The otter in the zoo ate $8\frac{2}{5}$ pounds of fish. The dolphin ate $2\frac{1}{2}$ times as much fish. How many pounds of fish did the dolphin eat?

Part 4	
a.	

b. Sally picked $1\frac{2}{3}$ baskets of grapes yesterday. She picked $3\frac{1}{5}$ times more baskets today. How many baskets of grapes did she pick today?

Independent Work

Part 5 Write an equation for each sentence.

a. The trailer was $\frac{7}{3}$ the length of the boat.

b. Josh was $\frac{3}{8}$ the age of his father.

c. The red car went $\frac{8}{5}$ as far as the blue car.

Part 5	
a.	

Part 6 Work each problem.

a. Jane was $9\frac{2}{5}$ centimeters taller than her brother. Her brother was $100\frac{1}{2}$ centimeters tall. How tall was Jane?

Part 6	
a.	

b. The first load the truck carried to the recycling center weighed $4\frac{1}{3}$ tons. The second load weighed $9\frac{2}{5}$ tons. How much did both loads weigh together? How much more did the second load weigh than the first load?

Connecting Math Concepts

Lesson 114

Part 7 Find the average for each line plot. Write the answer as a whole number or mixed number.

Part 7	
a.	

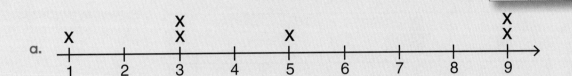

a.

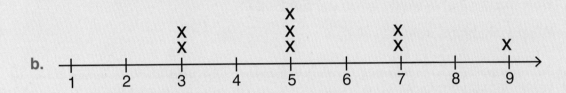

b.

Part 8 Work each problem.

a. They planted maple trees and pine trees. The ratio of pines to maples was 4 to 9. They planted 144 pines. How many maple trees did they plant? How many trees did they plant in all?

Part 8	
a.	

b. There were 7 gray rocks for every 3 white rocks in the jar. There was a total of 300 gray and white rocks in the jar. How many were gray? How many were white?

Part 9 Copy each problem. Simplify. Then multiply.

a. $\dfrac{360}{36}\left(\dfrac{12}{90}\right)$ 　　 b. $\dfrac{45}{9}\left(\dfrac{180}{9}\right)$ 　　 c. $\dfrac{20}{50}\left(\dfrac{15}{3}\right)$

Part 9	
a.	

Part 10 Work each item.

a. On Tuesday, the construction workers built $34\frac{1}{3}$ feet of sidewalk. In the afternoon they completed another 10 yards of sidewalk. How many feet of sidewalk did they complete during the day?

Part 10	
a.	

b. Bob's mother walked for $\frac{3}{4}$ hour in the morning and 45 minutes in the afternoon. How many hours did she walk that day?

Lesson 114

Part 11 Work each item.

- You want to make the beam balance by putting weights at one place on the right side.

Part 11	
a.	

a. How many weights would go on the right side?

b. Where would the weights go?

Part 12 Write the multiplication for the tens and ones. Then write the addition equation.

a. 5(73) b. 8(21)

Part 12	
a.	■(■)
	■(■) + ■(■)
	■ ■ + ■ = ■

Part 13 Work each item.

a. What's $\frac{1}{5}$ of $3\frac{2}{3}$ years?

b. What's $\frac{4}{7}$ of $1\frac{3}{8}$ days?

c. What's $\frac{3}{4}$ of $6\frac{1}{3}$ pounds?

Part 13	
a.	

Lesson 115

Part 1 Work each problem.

a. Bill was $\frac{4}{3}$ the height of his sister. His sister was 57 inches tall. How tall was Bill?

b. The white house is 48 years old. The gray house is $\frac{3}{4}$ as old as the white house. How old is the gray house?

c. The library had $\frac{9}{2}$ the number of books as the truck. There were 810 books in the library. How many books were in the truck?

Part 1	
a.	■ = ■ ■
	■

Independent Work

Part 2 Work each problem.

a. The horse drank $2\frac{2}{3}$ gallons of water a day. How much water did the horse drink in $5\frac{1}{2}$ days?

b. Each bag weighed 3 kilograms. How much did $4\frac{1}{2}$ bags weigh?

c. The business used $3\frac{1}{4}$ ounces of toner every hour. How much toner did the business use in $2\frac{2}{5}$ hours?

Part 2	
a.	

Part 3 Work each item. Use the measurement table if you need to.

a. Change 18 inches into yards.
b. Change 18 inches into feet.
c. Change 18 feet into yards.
d. Change 18 yards into feet.

Part 3	
a.	

Lesson 115

Part 4 Answer each question.

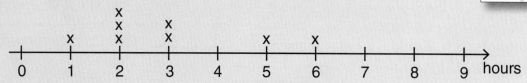

Time Students Spent on Homework

Part 4	
a.	

a. What was the average time spent on homework?

b. How many students spent less than 3 hours on homework?

c. How many students spent more than 3 hours on homework?

d. How many students spent 2 hours on homework?

e. What was the most frequent time students spent on homework?

f. How many students spent more than 6 hours on homework?

Part 5 Write the equation to show the fraction that equals each mixed number.

a. $2\frac{7}{9} = $ ■   b. $3\frac{3}{4} = $ ■   c. $1\frac{7}{14} = $ ■   d. $5\frac{3}{9} = $ ■

Part 5	
a.	

Part 6 Write an equation for each sentence.

a. The jar weighs $\frac{2}{3}$ as much as the box.

b. The new bridge is $\frac{7}{5}$ the length of the old bridge.

c. Elwood was $\frac{3}{11}$ the age of Norm.

Part 6	
a.	

Part 7 Copy each problem. Simplify. Then multiply.

a. $\frac{90}{6}\left(\frac{12}{270}\right)$ b. $\frac{21}{40}\left(\frac{200}{30}\right)$ c. $\frac{40}{6}\left(\frac{18}{120}\right)$

Part 7	
a.	

Connecting Math Concepts

Lesson 115

Part 8

Write the answer to each problem.

a. .045   (Multiply by 10.)

b. 4.02   (Divide by 100.)

c. 78.3   (Divide by 1000.)

d. 0.026   (Multiply by 1000.)

Part 8	
a.	

Part 9

Work each item.

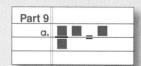

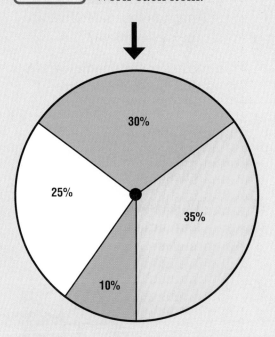

a. A person spins the wheel. When it stops, which color is most likely?

b. The person spins the wheel again. When it stops, which color is least likely?

c. If a person spins the wheel 20 times, about how many times would you expect the wheel to stop at white?

d. If a person spins the wheel 20 times, about how many times would you expect the wheel to stop at yellow?

Part 10

Work the problem. Write the answer as a mixed number with a unit name.

a. They drove to Alaska and then back home. The trip to Alaska took $2\frac{1}{4}$ days. The trip from Alaska took $3\frac{2}{3}$ days. What was the total time for both trips?

Part 10	
a.	

Part 11

Work each problem.

a. $\frac{5}{8}$ of the the children were boys. There were 150 girls. How many boys were there? How many children were there in all?

Part 11	
a.	

b. $\frac{2}{3}$ of the blankets were wet. There were 24 wet blankets. How many were not wet? How many blankets were there in all?

Lesson 115

Part 12   Write the names for the figures in a through e. Then answer the questions.

rhombus, parallelogram, quadrilateral, square, rectangle

Part 12	
a.	

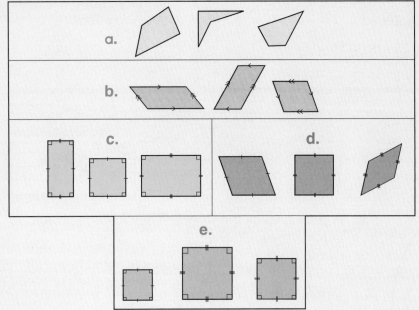

a.

b.

c.

d.

e.

f. What is any parallelogram that has four 90-degree angles and four sides the same length?

g. What is any parallelogram that has four sides that are the same length?

h. What is any parallelogram that has four 90-degree angles?

i. What is any quadrilateral that has two pairs of parallel sides?

j. What is any figure that has four straight sides and four angles?

Lesson

Part 1 Work each problem as mental math.

a. | Flowers of Different Colors a Person Pulled from a Box Without Looking |

yellow	2
red	7
blue	1

- How many flowers of each color would you expect the person to pull from the box if she took 40 trials?

---

b. | Marbles of Different Colors a Person Pulled from a Bag Without Looking |

white	8
pink	5
orange	7

- How many marbles of each color would you expect the person to pull from the bag if she took 120 trials?

Part 2 Work each problem.

a. Joe has 40 pounds of flour and wants to divide it equally into 32 bags. How many pounds of flour will be in each bag?

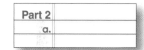

b. Lacy has 3 pounds of salad. She wants to make 4 equal servings. How much salad will be in each serving?

c. A strip of tape is 34 inches long. We divide it into 5 equal parts. How many inches long is each part?

Independent Work

Part 3 Work each item. Use the measurement table if you need to.

   a. Change 3.27 kilograms into grams.

   b. Change 56.1 centiliters into liters.

   c. Change 45.7 centimeters into millimeters.

Part 3	
a.	

Lesson 116

Part 4 Work the problem.

a. At the airport there are 5 large planes for every 9 small planes. There are 90 small planes. How many planes are there in all? How many large planes are there?

Part 4	
a.	

Part 5 Write the multiplication for the tens and ones. Then write the addition equation.

a. 7(16) b. 5(46)

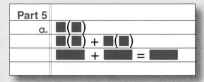

Part 6 Work each problem.

a. In the morning, the crew cleared $14\frac{5}{8}$ acres of swamp land. In the afternoon, the crew cleared $9\frac{1}{2}$ acres. How many acres of land did they clear that day?

Part 6	
a.	

b. In one year, Tom found $13\frac{1}{10}$ ounces of silver. His dad found $17\frac{1}{2}$ more ounces than Tom found. How much silver did his dad find?

Part 7 Work each item.

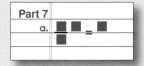

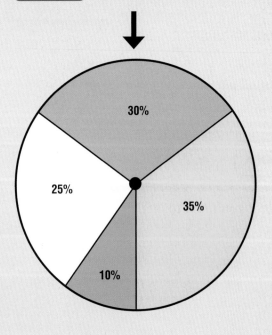

a. If a person spins the wheel 40 times, about how many times would you expect the wheel to stop at green?

b. A person will spin the wheel until it lands on white 5 times. About how many trials would you expect the person to take?

c. A person will spin the wheel until it lands on blue 6 times. About how many trials would you expect the person to take?

Lesson 116

Part 8 Work each item.

a. The bird weighed 18 ounces. The kitten weighed 2 pounds. How many ounces did the bird and kitten weigh together?

b. Dan ran 1500 meters on Monday and 2.5 kilometers on Tuesday. How many kilometers did he run on both days?

Part 8	
a.	

Part 9 Work each problem.

a. What's $\frac{3}{4}$ of 8 $\frac{1}{2}$ feet?

b. What's $\frac{9}{5}$ of 1 $\frac{1}{4}$ pounds?

c. Each day, Mary runs 1 $\frac{1}{2}$ miles. How far does Mary run in 5 days?

d. The smaller bag of mail weighed 10 $\frac{3}{5}$ pounds. The larger bag weighed 2 times as much as the smaller bag. How much did the larger bag weigh? How much did both bags weigh together?

e. What's $\frac{9}{2}$ of 10 $\frac{1}{3}$ cm?

f. What's $\frac{7}{8}$ of 8 days?

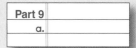

Part 9	
a.	

Part 10 Find the average for each line plot.

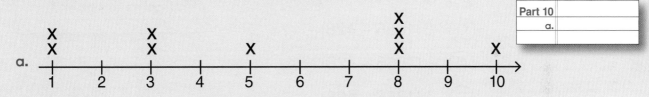

Part 10	
a.	

a.

b.

Part 11 Write an equation for each sentence.

a. The maple tree is $\frac{4}{5}$ the height of the fir tree.

b. Angie earned $\frac{7}{3}$ as much as Larry.

c. Chris was $\frac{4}{9}$ the age of Kammie.

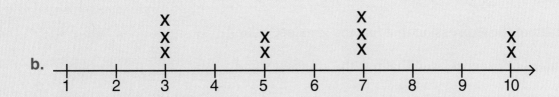

Part 11	
a.	

Lesson 117

Part 1 Work each problem.

- The fir board is 72 inches long. The maple board is $\frac{5}{3}$ the length of the fir board.

a. What's the length of the maple board?

b. What's the total length of both boards?

c. How much longer is the maple board than the fir board?

- In the study room, the number of books was $\frac{5}{2}$ the number of magazines. There were 95 books.

d. How many magazines were there?

e. How many fewer magazines than books were there?

f. What was the total number of books and magazines?

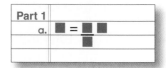

Part 2 Work each item.

$$\text{P} \quad 12\,(1856 - 428)$$

$$\text{Q} \quad 3\,(1856 - 428)$$

$$\text{R} \quad \tfrac{1}{4}\,(1856 - 428)$$

$$\text{S} \quad \tfrac{1}{2}\,(1856 - 428)$$

$$\text{T} \quad 4\,(1856 - 428)$$

Part 2	
a.	b.

a. Write the letter of the expression that has the greatest value.

b. Write the letter of the expression that has the smallest value.

c. Write the letters of the expressions that are more than $\frac{1}{2}\,(1856 - 428)$.

d. Write the letter of the expression that is 3 times greater than $4\,(1856 - 428)$.

> The quantity $(1856 - 428)$ is 1428.

e. Figure out what $3\,(1856 - 428)$ equals.

f. Figure out what $\frac{1}{2}\,(1856 - 428)$ equals.

Lesson 117

Part 3 Work each item.

a. What's $\frac{6}{5}$ of $2\frac{1}{8}$ miles?

b. What's $\frac{4}{9}$ of $1\frac{2}{3}$ yards?

c. What's $\frac{4}{10}$ of $3\frac{1}{6}$ tons?

d. What's $\frac{1}{2}$ of $\frac{1}{4}$ pound?

Part 3	
a.	

Part 4 Copy and work each problem.

a. $8\,6\,\overline{)\,5\,1\,5\,}$

b. $4\,2\,\overline{)\,2\,4\,1\,}$

Part 4	
a.	

Part 5 Work each problem.

a. The brown cat was $4\frac{1}{2}$ years old. The white cat was $2\frac{2}{3}$ times older than the brown cat. How old was the white cat?

Part 5	
a.	

b. Jane wrote at the average rate of 3 pages per hour. How many pages would she write in $4\frac{3}{4}$ hours?

c. The large box held $3\frac{1}{5}$ times as much material as the small box held. The small box held $2\frac{2}{9}$ cubic feet of material. How many cubic feet of material did the large box hold?

Part 6 Write the answer to each problem.

a. .001 (Divide by 10.)

b. 3.061 (Multiply by 100.)

c. 1.028 (Divide by 100.)

d. 3.03 (Multiply by 1000.)

Part 6	
a.	

Part 7 Work each problem. Write the answer as a mixed number with a unit name.

a. The white goat weighed $88\frac{2}{10}$ pounds. The spotted goat weighed $54\frac{1}{5}$ pounds. How much heavier was the white goat than the spotted goat?

Part 7	
a.	

Lesson 118

Part 1 Copy both fractions. Rewrite the fractions. Complete the statement with the sign >, <, or =.

a. $\dfrac{6}{10}$ ■ $\dfrac{2}{3}$

b. $\dfrac{9}{4}$ ■ $\dfrac{14}{6}$

c. $\dfrac{9}{15}$ ■ $\dfrac{6}{10}$

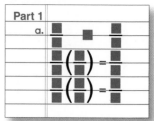

Part 2 Tell why each answer is impossible.

a. $\dfrac{7}{8} - \dfrac{1}{2} = \dfrac{8}{7}$

b. $\dfrac{2}{5} + \dfrac{7}{8} = \dfrac{9}{20}$

c. $\dfrac{10}{10} - \dfrac{4}{4} = \dfrac{6}{6}$

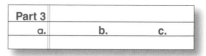

Part 3 Work each item.

- Ⓣ 2(1149 + 357)
- Ⓥ 3(1149 + 357)
- Ⓧ $\dfrac{4}{3}$(1149 + 357)
- Ⓨ $\dfrac{1}{2}$(1149 + 357)
- Ⓩ 6(1149 + 357)

Part 3		
a.	b.	c.

a. Write the letter of the expression that has the smallest value.

b. Write the letter of the expression that has the greatest value.

c. Write the letters of the expressions that are more than $\dfrac{4}{3}$(1149 + 357).

d. Write the letter of the expression that is 3 times greater than 2(1149 + 357).

e. Write the letter of the expression that is less than (1149 + 357).

f. Write the letters of all expressions that are more than (1149 + 357).

The quantity of (1149 + 357) is 1506.

g. Figure out what $\dfrac{4}{3}$(1149 + 357) equals.

h. Figure out what $\dfrac{1}{2}$(1149 + 357) equals.

Lesson 118

Independent Work

Part 4 Answer each question.

Number of Tickets Issued During the First 8 Days of May

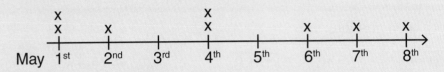

a. On which days were the largest number of tickets issued?

b. On which days were no tickets issued?

c. On how many days was one ticket issued?

d. On how many days was more than one ticket issued?

Part 4	
a.	

Part 5 Copy and work each problem.

a. $8\overline{)768}$ b. $7\overline{)394}$ c. $3\overline{)760}$ d. $9\overline{)934}$

Part 5	
a.	

Part 6 Work each problem.

a. Ernie threw the iron ball $65\frac{1}{4}$ feet. Tyrese threw the ball $77\frac{5}{8}$ feet. How much farther did Tyrese throw the ball than Ernie threw the ball?

b. Tim's backpack weighed $34\frac{2}{3}$ pounds. Jan's backpack weighed $55\frac{7}{8}$ pounds. How much did the two backpacks weigh together?

Part 6	
a.	

Part 7 Copy each problem. Simplify. Then multiply.

a. $\frac{90}{30}\left(\frac{28}{10}\right)$ b. $\frac{8}{7}\left(\frac{35}{12}\right)$ c. $\frac{16}{50}\left(\frac{300}{24}\right)$

Part 7	
a.	

Part 8 Write the equation to show the fraction that equals each mixed number.

a. $2\frac{5}{8} = \blacksquare$ b. $8\frac{1}{2} = \blacksquare$ c. $7\frac{4}{9} = \blacksquare$

Part 8	
a.	

Lesson 118

Part 9 Work each item.

a. What's $\frac{1}{9}$ of $2\frac{3}{5}$ months?

b. What's $\frac{8}{3}$ of $1\frac{1}{2}$ pounds?

c. What's $\frac{5}{8}$ of 9 days?

d. What's $\frac{7}{10}$ of $1\frac{2}{3}$ ounces?

Part 9	
a.	

Part 10 Copy and work each problem.

a. $50\overline{)360}$

b. $17\overline{)148}$

Part 10	
a.	

Part 11 Work each problem.

a. The chair cost $\frac{1}{3}$ as much as the table. The table cost $96. How much did the chair cost? How much did the table and one chair cost? How much more was the cost of the table than the cost of the chair?

Part 11	
a.	

b. The car was $\frac{3}{4}$ the length of the boat. The car was 18 feet long. How long was the boat? How much longer was the boat than the car? What's the total length of both vehicles?

c. John's house was $\frac{7}{5}$ the age of Brenda's house. John's house was 35 years old. How old was Brenda's house? How much older was John's house than Brenda's house?

Lesson 119

Part 1

Copy both fractions. Rewrite fractions if you need to. Complete the statement with the sign >, <, or =.

a. $\dfrac{7}{5}$ ▪ $\dfrac{13}{10}$ b. $\dfrac{7}{4}$ ▪ $\dfrac{8}{5}$ c. $\dfrac{3}{6}$ ▪ $\dfrac{5}{9}$ d. $\dfrac{3}{4}$ ▪ $\dfrac{9}{12}$

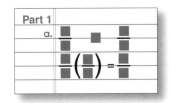

Part 2

For each item, write **possible** or **impossible.** If the answer is impossible, tell why.

a. $\dfrac{3}{5} + \dfrac{2}{20} = \dfrac{7}{10}$ b. $\dfrac{3}{5} - \dfrac{2}{20} = \dfrac{5}{10}$

c. $\dfrac{3}{5} + \dfrac{6}{10} = \dfrac{18}{20}$ d. $\dfrac{3}{5} - \dfrac{2}{20} = \dfrac{11}{10}$

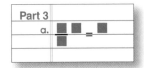

Part 3

Write an equation for each sentence.

a. $\dfrac{4}{7}$ of the pens cost 88 dollars.

b. $\dfrac{2}{3}$ of the gravel weighed 502 pounds.

c. $\dfrac{6}{7}$ of the water was $\dfrac{5}{8}$ gallon.

d. $\dfrac{3}{10}$ of the classroom was 500 square feet.

Connecting Math Concepts

Lesson 119

Independent Work

Part 4 Work each problem.

a. $\frac{3}{5}$ of the jeans cost 240 dollars. How much do all the jeans cost?

b. Henry is $\frac{5}{4}$ the age of his barn. His barn is 60 years old. How old is Henry?

c. The new car is $\frac{2}{3}$ the weight of the used car. The new car weighs 2800 pounds. What is the weight of the used car? How much do the two cars weigh together?

d. The turtle was $\frac{7}{5}$ the age of the parrot. The parrot was 40 years old. How old was the turtle? How much older was the turtle than the parrot?

e. $\frac{3}{4}$ of the eggs weigh 24 ounces. What do all the eggs weigh?

Part 4	
a.	

Part 5 Write the multiplication for the tens and ones. Then write the addition equation.

a. 3(76)   b. 7(45)

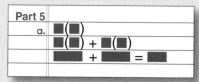

Part 5	
a.	

Part 6 Work each problem.

a. The team drank $2\frac{3}{4}$ gallons of water during Wednesday's practice. The team drank $3\frac{11}{12}$ gallons of water during the Thursday practice. How much water did the team drink on both days?

b. Lisa's sister was $2\frac{3}{4}$ years older than Lisa. Her sister was $16\frac{1}{4}$ years old. How old was Lisa?

Part 6	
a.	

Part 7 Copy and work each problem.

a. $7\overline{)503}$ b. $4\overline{)364}$ c. $32\overline{)187}$

Part 7	
a.	

Part 8 Write the equation to show the fraction that equals each mixed number.

a. $3\frac{4}{5} = $ ■   b. $10\frac{2}{13} = $ ■   c. $4\frac{4}{9} = $ ■

Part 8	
a.	

Lesson 120

For each item, write **possible** or **impossible.** If the answer is impossible, tell why.

a. $\dfrac{1}{2} - \dfrac{3}{8} = \dfrac{4}{8}$

b. $\dfrac{7}{7} - \dfrac{9}{14} = \dfrac{10}{28}$

c. $\dfrac{9}{10} - \dfrac{1}{5} = \dfrac{22}{20}$

d. $\dfrac{2}{5} + \dfrac{9}{15} = \dfrac{30}{30}$

Part 1	
a.	

Independent Work

Part 2 Find the surface area of the rectangular prism.

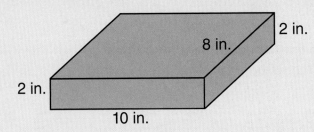

2 in.

8 in.

2 in.

10 in.

Part 2	
a.	SA = 2(■ + ■ + ■)

Part 3 Work each item.

a. What's $\dfrac{4}{5}$ of $12\dfrac{1}{2}$ pounds?

b. What's $\dfrac{1}{2}$ of $2\dfrac{3}{4}$ tons?

c. What's $\dfrac{5}{8}$ of $1\dfrac{6}{10}$ inches?

Part 3	
a.	

Part 4 Copy and work each problem.

a. $53\overline{)689}$

b. $62\overline{)259}$

Part 4	
a.	

Lesson 120

Independent Work

Part 5 | Work each problem.

a. The couch weighed $\frac{10}{3}$ as much as the table. The table weighed 18 pounds. How much did the couch weigh?

Part 5	
a.	

b. The car was $\frac{4}{5}$ the length of the boat. The car was 24 feet long. What was the length of the boat?

c. The distance to town A is $\frac{6}{5}$ the distance to town B. The distance to town B is 30 miles. What's the distance to town A?

d. The kettle held $\frac{3}{8}$ as much as the barrel. The kettle held 4 gallons. How much did the barrel hold?

Part 6 | Copy both fractions. Rewrite fractions if you need to. Complete the statement with the sign >, <, or =.

a. $\frac{5}{4}$ ■ $\frac{15}{12}$ b. $\frac{11}{4}$ ■ $\frac{10}{3}$ c. $\frac{9}{10}$ ■ $\frac{4}{5}$ d. $\frac{3}{6}$ ■ $\frac{5}{9}$

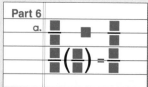

Part 7 | Write the names for the figures in a through e. Then answer the questions.

rectangle, rhombus, square, quadrilateral, parallelogram

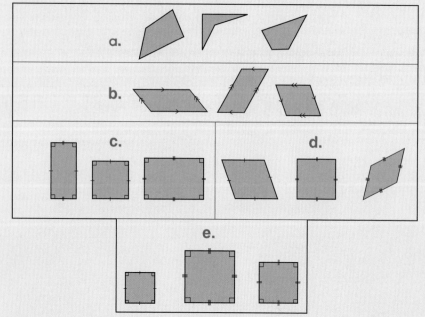

Part 7	
a.	

f. What is any rectangle with all sides the same length?

g. What is any quadrilateral with two pairs of parallel sides?

h. What is any parallelogram with four 90-degree angles?

i. What is any parallelogram with all the sides the same?

j. What is any figure with 4 straight sides and 4 angles?

Connecting Math Concepts

Lesson 120

Part 8 Work each problem.

a. The distance to Harry's house is $4\frac{2}{3}$ times the distance to Jan's house. The distance to Jan's house is $2\frac{1}{6}$ miles. What's the distance to Harry's house?

Part 8	
a.	

b. The turtle weighed $9\frac{1}{4}$ pounds. The snake weighed $\frac{2}{5}$ as much as the turtle. How much did the snake weigh?

c. The new tent is $1\frac{1}{2}$ times the height of the old tent. The old tent is $5\frac{1}{2}$ feet tall. How tall is the new tent?

Part 9 Answer each question.

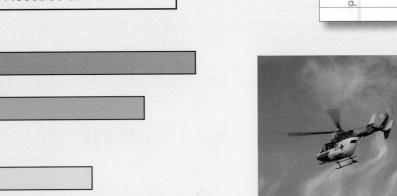

Mountain Rescues at Mount Hood

Part 9	
a.	

a. How many rescues occurred during the years 1995 to 2002?

b. What is the average number of rescues each year?

c. Which year had the most rescues?

d. How many years had fewer than 2 rescues?

e. How many years had more than 2 rescues?

Level F Correlation to Grade 5
Common Core State Standards for Mathematics

Operations and Algebraic Thinking (5.OA)

Write and interpret numerical expressions.

1. Use parentheses, brackets, or braces in numerical expressions, and evaluate expressions with these symbols.

Lessons	WB: 16–23, 35, 37, 39–41, 69, 90, 103, 104, 108, 115–117 TB: 22–79, 83–120 Student Practice Software: Block 2 Activity 6 and Block 3 Activity 6

Operations and Algebraic Thinking (5.OA)

Write and interpret numerical expressions.

2. Write simple expressions that record calculations with numbers, and interpret numerical expressions without evaluating them. *For example, express the calculation "add 8 and 7, then multiply by 2" as 2 × (8 + 7). Recognize that 3 × (18932 + 921) is three times as large as 18932 + 921, without having to calculate the indicated sum or product.*

Lessons	WB: 119, 120 TB: 46–48, 88–93, 117, 118

Operations and Algebraic Thinking (5.OA)

Write and interpret numerical expressions.

*2.1 Express a whole number in the range 2-50 as a product of its prime factors. For example, find the prime factors of 24 and express 24 as 2 x 2 x 2 x 3.

Lessons	WB: 51, 52, 55, 56 TB: 53, 54, 56–59, 61, 74, 77

*Denotes California-only content.

Operations and Algebraic Thinking (5.OA)

Analyze patterns and relationships.

3. Generate two numerical patterns using two given rules. Identify apparent relationships between corresponding terms. Form ordered pairs consisting of corresponding terms from the two patterns, and graph the ordered pairs on a coordinate plane. *For example, given the rule "Add 3" and the starting number 0, and given the rule "Add 6" and the starting number 0, generate terms in the resulting sequences, and observe that the terms in one sequence are twice the corresponding terms in the other sequence. Explain informally why this is so.*

Lessons	WB: 118, 119, 120

Number and Operations in Base Ten (5.NBT)

Understand the place value system.

1. Recognize that in a multi-digit number, a digit in one place represents 10 times as much as it represents in the place to its right and 1/10 of what it represents in the place to its left.

Lessons	WB: 17, 64, 65, 70, 72, 81–84, 93, 95–97, 101, 102 TB: 63, 67–72, 75, 85–89, 98, 101, 103–106, 108–113, 115, 117

Number and Operations in Base Ten (5.NBT)

Understand the place value system.

2. Explain patterns in the number of zeros of the product when multiplying a number by powers of 10, and explain patterns in the placement of the decimal point when a decimal is multiplied or divided by a power of 10. Use whole-number exponents to denote powers of 10.

Lessons	WB: 64, 65, 70, 72, 81–84, 93, 95–97, 101, 102 TB: 63, 67–72, 85–89, 98, 101, 103–106, 108–113, 115, 117

Number and Operations in Base Ten (5.NBT)

Understand the place value system.

3. Read, write, and compare decimals to thousandths.
   a. Read and write decimals to thousandths using base-ten numerals, number names, and expanded form, e.g., $347.392 = 3 \times 100 + 4 \times 10 + 7 \times 1 + 3 \times (1/10) + 9 \times (1/100) + 2 \times (1/1000)$.
   b. Compare two decimals to thousandths based on meanings of the digits in each place, using >, =, and < symbols to record the results of comparisons.

Lessons	WB: 23–25, 28, 29, 32–37, 39, 43, 45–47, 54, 57–65, 72, 74–85, 88, 92, 93, 95, 97, 98, 100 TB: 26, 27, 29, 30–32, 67–70, 102–105, 107 Student Practice Software: Block 2 Activities 2 and 4 and Block 4 Activity 4

Number and Operations in Base Ten (5.NBT)

Understand the place value system.

4. Use place value understanding to round decimals to any place.

Lessons	WB: 83 TB: 67–76, 79, 81, 85, 87 Student Practice Software: Block 4 Activity 3

Number and Operations in Base Ten (5.NBT)

Perform operations with multi-digit whole numbers and with decimals to hundredths.

5. Fluently multiply with multi-digit whole numbers using the standard algorithm.

Lessons	WB: 1–20, 33, 34, 36–38, 44, 45, 48, 51, 54, 57, 94, 96–99, 103, 104, 107, 115 TB: 1, 17–19, 21–24, 26–43, 45–47, 49–60, 66–68, 71, 75, 84–114, 116–120 Student Practice Software: Block 1 Activity 6

Number and Operations in Base Ten (5.NBT)

Perform operations with multi-digit whole numbers and with decimals to hundredths.

6. Find whole-number quotients of whole numbers with up to four-digit dividends and two-digit divisors, using strategies based on place value, the properties of operations, and/or the relationship between multiplication and division. Illustrate and explain the calculation by using equations, rectangular arrays, and/or area models.

Lessons	WB: 1–20, 22, 24, 26, 28–30, 35, 66–70, 81, 95, 96, 104, 107, 108 TB: 14, 16, 18, 19, 21–28, 30, 31, 34, 36–40, 44, 46–57, 59, 61, 62, 65, 73, 75, 77, 82–91, 93, 95–102, 105, 106, 109–111, 114–120 Student Practice Software: Block 1 Activity 5 and Block 3 Activity 5

Number and Operations in Base Ten (5.NBT)

Perform operations with multi-digit whole numbers and with decimals to hundredths.

7. Add, subtract, multiply, and divide decimals to hundredths, using concrete models or drawings and strategies based on place value, properties of operations, and/or the relationship between addition and subtraction; relate the strategy to a written method and explain the reasoning used.

Lessons	WB: 36, 37, 40, 42, 48, 50, 61–63, 66–72, 74, 76, 79–81, 84, 93, 95–97, 102, 113, 114, 118–120 TB: 38, 40, 41, 43–56, 58–61, 63–68, 71, 73–75, 77–82, 84–87, 89, 91, 93, 94, 97–99, 101, 103–105, 108, 109, 111 Student Practice Software: Block 2 Activity 3, Block 3 Activity 1, and Block 4 Activity 5

Number and Operations—Fractions (5.NF)

Use equivalent fractions as a strategy to add and subtract fractions.

1. Add and subtract fractions with unlike denominators (including mixed numbers) by replacing given fractions with equivalent fractions in such a way as to produce an equivalent sum or difference of fractions with like denominators. *For example, 2/3 + 5/4 = 8/12 + 15/12 = 23/12. (In general, a/b + c/d = (ad + bc)/bd.)*

Lessons	WB: 16–23, 91, 92 TB: 24–31, 33, 35, 36, 38, 40, 41, 43, 46, 48, 49, 50, 52, 53, 56, 78, 80, 84, 92–99, 101–104, 106–111, 113–119 Student Practice Software: Block 1 Activity 4 and Block 5 Activity 2

Number and Operations—Fractions (5.NF)

Use equivalent fractions as a strategy to add and subtract fractions.

2. Solve word problems involving addition and subtraction of fractions referring to the same whole, including cases of unlike denominators, e.g., by using visual fraction models or equations to represent the problem. Use benchmark fractions and number sense of fractions to estimate mentally and assess the reasonableness of answers. *For example, recognize an incorrect result 2/5 + 1/2 = 3/7, by observing that 3/7 < 1/2.*

Lessons	TB: 45, 46, 48–50, 52, 53, 56, 78, 80–87, 90, 92, 96–98, 100–111, 113–120 Student Practice Software: Block 5 Activity 2

Number and Operations—Fractions (5.NF)

Apply and extend previous understandings of multiplication and division to multiply and divide fractions.

3. Interpret a fraction as division of the numerator by the denominator ($a/b = a \div b$). Solve word problems involving division of whole numbers leading to answers in the form of fractions or mixed numbers, e.g., by using visual fraction models or equations to represent the problem. *For example, interpret 3/4 as the result of dividing 3 by 4, noting that 3/4 multiplied by 4 equals 3, and that when 3 wholes are shared equally among 4 people each person has a share of size 3/4. If 9 people want to share a 50-pound sack of rice equally by weight, how many pounds of rice should each person get? Between what two whole numbers does your answer lie?*

Lessons	WB: 10–12, 16–22, 24, 35, 39, 46, 47, 50, 75, 76, 116
	TB: 8, 9, 13–15, 17–19, 21–26, 28, 30–34, 37, 39, 42, 44, 46–50, 53, 59, 60, 89, 94, 114, 116

Number and Operations—Fractions (5.NF)

Apply and extend previous understandings of multiplication and division to multiply and divide fractions.

4. Apply and extend previous understandings of multiplication to multiply a fraction or whole number by a fraction.
   a. Interpret the product (a/b) × q as a parts of a partition of q into b equal parts; equivalently, as the result of a sequence of operations $a \times q \div b$. For example, use a visual fraction model to show (2/3) × 4 = 8/3, and create a story context for this equation. Do the same with (2/3) × (4/5) = 8/15. (In general, (a/b) × (c/d) = ac/bd.)
   b. Find the area of a rectangle with fractional side lengths by tiling it with unit squares of the appropriate unit fraction side lengths, and show that the area is the same as would be found by multiplying the side lengths. Multiply fractional side lengths to find areas of rectangles, and represent fraction products as rectangular areas.

Lessons	WB: 11–21, 35, 39–41, 60, 89–91, 94, 99, 100, 103, 107, 108, 115–117, 119, 120
	TB: 9–15, 21–34, 36–56, 58–120
	Student Practice Software: Block 2 Activity 6, Block 3 Activities 3 and 6, and Block 5 Activity 4

Number and Operations—Fractions (5.NF)

Apply and extend previous understandings of multiplication and division to multiply and divide fractions.

5. Interpret multiplication as scaling (resizing), by:
   a. Comparing the size of a product to the size of one factor on the basis of the size of the other factor, without performing the indicated multiplication.
   b. Explaining why multiplying a given number by a fraction greater than 1 results in a product greater than the given number (recognizing multiplication by whole numbers greater than 1 as a familiar case); explaining why multiplying a given number by a fraction less than 1 results in a product smaller than the given number; and relating the principle of fraction equivalence $a/b = (n \times a)/(n \times b)$ to the effect of multiplying a/b by 1.

Lessons	WB: 11–21, 35, 59–64, 66, 69–76, 79, 81, 82, 84, 89–91, 94, 96, 114–117
	TB: 22–41, 43, 45–47, 49, 51–56, 58–63, 65, 67, 68, 73, 74, 76–80, 83, 85, 87, 91–93, 95, 97, 98, 100, 103, 105, 118–120
	Student Practice Software: Block 2 Activity 6 and Block 4 Activity 1

Number and Operations—Fractions (5.NF)

Apply and extend previous understandings of multiplication and division to multiply and divide fractions.

6. Solve real world problems involving multiplication of fractions and mixed numbers, e.g., by using visual fraction models or equations to represent the problem.

Lessons	TB: 49–55, 57, 59, 61, 62, 64, 66, 68, 70–87, 90, 91, 93–100, 103, 104, 106–120 Student Practice Software: Block 3 Activity 6 and Block 5 Activity 4

Number and Operations—Fractions (5.NF)

Apply and extend previous understandings of multiplication and division to multiply and divide fractions.

7. Apply and extend previous understandings of division to divide unit fractions by whole numbers and whole numbers by unit fractions.
   a. Interpret division of a unit fraction by a non-zero whole number, and compute such quotients. *For example, create a story context for (1/3) ÷ 4, and use a visual fraction model to show the quotient. Use the relationship between multiplication and division to explain that (1/3) ÷ 4 = 1/12 because (1/12) × 4 = 1/3.*
   b. Interpret division of a whole number by a unit fraction, and compute such quotients. *For example, create a story context for 4 ÷ (1/5), and use a visual fraction model to show the quotient. Use the relationship between multiplication and division to explain that 4 ÷ (1/5) = 20 because 20 × (1/5) = 4.*
   c. Solve real world problems involving division of unit fractions by non-zero whole numbers and division of whole numbers by unit fractions, e.g., by using visual fraction models and equations to represent the problem. *For example, how much chocolate will each person get if 3 people share 1/2 lb of chocolate equally? How many 1/3-cup servings are in 2 cups of raisins?*

Lessons	WB: 73, 75, 76, 79–82, 115–117 TB: 77–81, 83–98, 113, 114, 116, 117

Measurement and Data (5.MD)

Convert like measurement units within a given measurement system.

1. Convert among different-sized standard measurement units within a given measurement system (e.g., convert 5 cm to 0.05 m), and use these conversions in solving multi-step, real world problems.

Lessons	TB: 91, 92, 95–105, 108–116 Student Practice Software: Block 5 Activity 1

Measurement and Data (5.MD)

Represent and interpret data.

2. Make a line plot to display a data set of measurements in fractions of a unit (1/2, 1/4, 1/8). Use operations on fractions for this grade to solve problems involving information presented in line plots. *For example, given different measurements of liquid in identical beakers, find the amount of liquid each beaker would contain if the total amount in all the beakers were redistributed equally.*

Lessons	WB: 112, 115, 116, 117, 119, 120 TB: 113, 114

Measurement and Data (5.MD)

Geometric measurement: understand concepts of volume and relate volume to multiplication and to addition.

3. Recognize volume as an attribute of solid figures and understand concepts of volume measurement.
   a. A cube with side length 1 unit, called a "unit cube," is said to have "one cubic unit" of volume, and can be used to measure volume.
   b. A solid figure which can be packed without gaps or overlaps using *n* unit cubes is said to have a volume of *n* cubic units.

Lessons	WB: 69
	Student Practice Software: Block 4 Activity 6 and Block 5 Activities 3 and 6

Measurement and Data (5.MD)

Geometric measurement: understand concepts of volume and relate volume to multiplication and to addition.

4. Measure volumes by counting unit cubes, using cubic cm, cubic in, cubic ft, and improvised units.

Lessons	WB: 69
	Student Practice Software: Block 4 Activity 6 and Block 5 Activities 3 and 6

Measurement and Data (5.MD)

Geometric measurement: understand concepts of volume and relate volume to multiplication and to addition.

5. Relate volume to the operations of multiplication and addition and solve real world and mathematical problems involving volume.
   a. Find the volume of a right rectangular prism with whole-number side lengths by packing it with unit cubes, and show that the volume is the same as would be found by multiplying the edge lengths, equivalently by multiplying the height by the area of the base. Represent threefold whole-number products as volumes, e.g., to represent the associative property of multiplication.
   b. Apply the formulas $V = l \times w \times h$ and $V = b \times h$ for rectangular prisms to find volumes of right rectangular prisms with whole-number edge lengths in the context of solving real world and mathematical problems.
   c. Recognize volume as additive. Find volumes of solid figures composed of two non-overlapping right rectangular prisms by adding the volumes of the non-overlapping parts, applying this technique to solve real world problems.

Lessons	WB: 69, 70
	TB: 71–74, 77, 80, 82, 84, 86, 89, 93–99, 101, 102

Geometry (5.G)

Graph points on the coordinate plane to solve real-world and mathematical problems.

1. Use a pair of perpendicular number lines, called axes, to define a coordinate system, with the intersection of the lines (the origin) arranged to coincide with the 0 on each line and a given point in the plane located by using an ordered pair of numbers, called its coordinates. Understand that the first number indicates how far to travel from the origin in the direction of one axis, and the second number indicates how far to travel in the direction of the second axis, with the convention that the names of the two axes and the coordinates correspond (e.g., *x*-axis and *x*-coordinate, *y*-axis and *y*-coordinate).

Lessons	WB: 24–34, 36, 42, 48, 54, 55, 57, 59, 62, 102, 104, 106, 111, 118–120 TB: 21–26, 33–37, 39, 45 Student Practice Software: Block 2 Activity 5

Geometry (5.G)

Graph points on the coordinate plane to solve real-world and mathematical problems.

2. Represent real world and mathematical problems by graphing points in the first quadrant of the coordinate plane, and interpret coordinate values of points in the context of the situation.

Lessons	WB: 29–34, 36, 39, 42, 48, 54, 55, 57, 59, 62, 65–67, 69, 70, 72–77, 79, 82–84, 86–88, 91, 94, 96, 99, 101, 105, 109, 111, 118–120 Student Practice Software: Block 2 Activity 5

Geometry (5.G)

Classify two-dimensional figures into categories based on their properties.

3. Understand that attributes belonging to a category of two-dimensional figures also belong to all subcategories of that category. *For example, all rectangles have four right angles and squares are rectangles, so all squares have four right angles.*

Lessons	WB: 109 TB: 110, 111, 112, 115, 120

Geometry (5.G)

Classify two-dimensional figures into categories based on their properties.

4. Classify two-dimensional figures in a hierarchy based on properties.

Lessons	WB: 109 TB: 110, 111, 112, 115, 120

Standards for Mathematical Practice

Connecting Math Concepts addresses all of the Standards for Mathematical Practice throughout the program. What follows are examples of how individual standards are addressed in this level.

1. Make sense of problems and persevere in solving them.

Word Problems: Basic Operations and Algebraic Translation (Lessons 1–117, intermittently): Students learn to identify specific types of word problems (e.g., comparison, unit conversion) involving whole numbers, fractions, and mixed numbers, and learn strategies to solve the problems based on the specific problem types.

2. Reason abstractly and quantitatively.

Decimal Operations (Lessons 116–120) and Fraction Operations (Lessons 114–120): Students start working with decimals and fractions within the first 20 lessons of *CMC Level F.* They learn to add, subtract, multiply, and divide. Toward the end of the program, they relate pictorial representations of fractions and decimals to addition and subtraction of fractions and decimals, attending to them as quantities with parts that can be manipulated.

3. Construct viable arguments and critique the reasoning of others.

Fraction Operations (Lessons 87–94): Students use their knowledge of fraction multiplication to reason and explain how they know facts about fraction multiplication problems without completing any calculation. For example, "If we multiply A/B by 2/3, will we end up with a fraction that is more than, less than, or equal to A/B? How do you know?"

4. Model with mathematics.

Coordinate System (Lessons 20–34 and 63–75): Students learn about the coordinate system, including how to plot points and lines on the *x* and *y* axes. Later, they plot data points and draw lines to model information given in word problems and use that data to answer questions.

5. Use appropriate tools strategically.

Throughout the program (Lessons 1–120) students use pencils, workbooks, lined paper, and textbooks to complete their work. They use rulers to draw lines on the coordinate system. Students also use the computer to access the Practice Software where they apply the skills they learn in the lessons.

6. Attend to precision.

Geometry (Lessons 40–59, 73–85, 105–111): When finding area, perimeter, and volume, students attend to the units and respond verbally and in written answers with the correct unit. They include units in answers to word problems that involve specific units.

7. Look for and make use of structure.

Inverse Operations (Lessons 103–112): In Lesson 103, students learn about the distributive property of multiplication. Starting in Lesson 106, they learn to apply the distributive property to area diagrams that show the product for the tens and the product for the ones (e.g., 4(32) = 4(30 + 2) = 120 + 8. Students relate the same diagram to the corresponding division problem (128 ÷ 4), and verify that the tens digit and ones digit in the quotient correspond to the tens and ones shown in each row of the diagram.

8. Look for and express regularity in repeated reasoning.

Mental Math (Lessons 10–45): Students apply understandings about different patterns to complete calculations mentally. The repeated reasoning is apparent in their ability to complete sequences of problems that apply the same pattern or numbers (e.g., 6 ÷ 2 followed by 60 ÷ 2, 60 ÷ 3, and 600 ÷ 3) and do so without mistakes.

Photo Credits